P9-CMO-571

FIELD THEORY AND ITS CLASSICAL PROBLEMS

By

CHARLES ROBERT HADLOCK

THE

CARUS MATHEMATICAL MONOGRAPHS

Published by

THE MATHEMATICAL ASSOCIATION OF AMERICA

THE CARUS MATHEMATICAL MONOGRAPHS are an expression of the desire of Mrs. Mary Hegeler Carus, and of her son, Dr. Edward H. Carus, to contribute to the dissemination of mathematical knowledge by making accessible at nominal cost a series of expository presentations of the best thoughts and keenest researches in pure and applied mathematics. The publication of the first four of these monographs was made possible by a notable gift to the Mathematical Association of America by Mrs. Carus as sole trustee of the Edward C. Hegeler Trust Fund. The sales from these have resulted in the Carus Monograph Fund, and the Mathematical Association has used this as a revolving book fund to publish the succeeding monographs.

The expositions of mathematical subjects which the monographs contain are set forth in a manner comprehensible not only to teachers and students specializing in mathematics, but also to scientific workers in other fields, and especially to the wide circle of thoughtful people who, having a moderate acquaintance with elementary mathematics, wish to extend their knowledge without prolonged and critical study of the mathematical journals and treatises. The scope of this series includes also historical and biographical monographs.

The following monographs have been published:

No. 1. Calculus of Variations, by G. A. BLISS

No. 2. Analytic Functions of a Complex Variable, by D. R. CURTISS

No. 3. Mathematical Statistics, by H. L. RIETZ

No. 4. Projective Geometry, by J. W. YOUNG

No. 5. A History of Mathematics in America before 1900, by D. E. SMITH and JEKUTHIEL GINSBURG (out of print)

No. 6. Fourier Series and Orthogonal Polynomials, by DUNHAM JACKSON

No. 7. Vectors and Matrices, by C. C. MACDUFFEE

No. 8. Rings and Ideals, by N. H. McCOY

No. 9. The Theory of Algebraic Numbers, Second edition, by HARRY POLLARD and HAROLD G. DIAMOND

No. 10. The Arithmetic Theory of Quadratic Forms, by B. W. JONES

No. 11. Irrational Numbers, by IVAN NIVEN

No. 12. Statistical Independence in Probability, Analysis and Number Theory, by MARK KAC

No. 13. A Primer of Real Functions, by RALPH P. BOAS, JR.

No. 14. Combinatorial Mathematics, by HERBERT JOHN RYSER

No. 15. Noncommutative Rings, by I. N. HERSTEIN

No. 16. Dedekind Sums, by HANS RADEMACHER and EMIL GROSSWALD

No. 17. The Schwarz Function and its Applications, by PHILIP J. DAVIS

No. 18. Celestial Mechanics, by HARRY POLLARD

No. 19. Field Theory and Its Classical Problems, by CHARLES ROBERT HADLOCK

The Carus Mathematical Monographs

NUMBER NINETEEN

FIELD THEORY AND ITS CLASSICAL PROBLEMS

By

CHARLES ROBERT HADLOCK

Arthur D. Little, Inc.

Published and Distributed by

THE MATHEMATICAL ASSOCIATION OF AMERICA

Library of Congress Catalog Card Number 78-71937

Complete Set ISBN 0-88385-000-1
Vol. 19 ISBN 0-88385-020-6

Printed in the United States of America

Current printing (last digit):

10 9 8 7 6 5 4 3 2 1

This book is dedicated to the memory of Charles William Hadlock, Grandfather.

PREFACE

I wrote this book for myself.

I wanted to piece together carefully my own path through Galois Theory, a subject whose mathematical centrality and beauty I had often glimpsed, but one which I had never properly organized in my own mind. I wanted to start with simple, interesting questions and solve them as quickly and directly as possible. If related interesting questions arose along the way, I would deal with them too, but only if they seemed irresistible. I wanted to avoid generality for its own sake, and, as far as practicable, even generality that could only be appreciated in retrospect. Thus, I approached this project as an inquirer rather than as an expert, and I hope to share some of the sense of discovery and excitement I experienced. There is great mathematics here.

In particular, the book presents an exposition of those portions of classical field theory which are encountered in the solution of the famous geometric construction problems of antiquity and the problem of solving polynomial equations by radicals. Some time ago much of this material was covered in undergraduate courses in the 'theory of equations'. Paradoxically, as the theory matured and became more elegant, it also moved higher into the curriculum, so that nowadays it is not uncommon for it first to be encountered on the graduate level. It seems to me that this is most unfortunate, for this important and beautiful area of mathematics deserves to be studied early. It can then lend perspective and motivation to the later abstract study of mathematical structures.

Simple mathematical questions and problems can be fascinating; theorems can often appear quite boring—unless they help us to deal with problems we are already thinking about. In this spirit, I have tried to develop the subject along the lines of various natural questions, usually stated at the beginnings of sections. Some of the mathematical gems encountered along the way relate to: angle trisection, duplication of the cube, squaring the circle, the transcendence of e and of π, the construction of regular polygons, the irreducibility of the nth cyclotomic polynomial, the solution of polynomial equations by radicals, Hilbert's irreducibility theorem, polynomials over the rationals with symmetric groups, the number of automorphisms of the field of complex numbers, and more.

This book differs from the standard textbooks in several respects. First, the prerequisites are quite modest: basic calculus, linear algebra (including matrices, determinants, eigenvalues, and vector spaces), and the ability to follow precise mathematical arguments. No previous knowledge of groups, fields, or abstract algebra is needed. Second, I work entirely with subfields of the field of complex numbers. The richness of the subject is almost fully present here, and the concrete setting makes it easier to grapple with and appreciate difficult material. Third, milestone developments appear almost immediately, in the first chapter, before we start to build up much theory. This provides momentum, and it also underlines the way the later theory represents the development of more powerful methods. Indeed, a number of problems are solved in several ways as we proceed. Fourth, the mathematics is 'interdisciplinary'. I have the luxury of not needing to follow a certain syllabus in abstract algebra, say, and so I develop and use whatever I need to solve a problem. Thus honest use is made of calculus, linear algebra, number theory, and algebraic geometry, for example, although without exceeding the prerequisites for the book. Fifth, there are reasonably complete solutions to all 176 problems. Too often have I turned to reference books only to find my own question given as a problem; it is often impractical to take the time to work such things out. Sixth, each chapter closes with an annotated bibliography, which should enable the interested reader to get quickly to the original sources or to supplementary material. In all, I hope this book may serve a wide audience: bright undergraduates, even at the sophomore-junior level, where I used the first three chapters for a one-semester course, prospective and current high school teachers, non-mathematicians with an interest in mathematics, as well as, yes, graduate students and professional mathematicians, some of whom have already welcomed the manuscript in the spirit of its unofficial title: "What You Always Wanted to Know about Galois Theory (but were afraid to ask)".

In order to maintain a sense of continuity, the thirty-seven theorems are numbered consecutively throughout the text. Lemmas and corollaries assume the number of the theorem they immediately precede or follow, respectively. There are certain instances where I use standard notions from calculus even though it is possible to work with formally identical counterparts that do not depend on calculus. For example, I treat polynomials as functions and their derivatives as these occur in calculus. I do not treat abstract groups or rings. It is true that in a few instances such concepts could be used to effect economies or more elegant arguments. It is also true that the similarities in a number of calculations point strongly and clearly at fundamental abstract notions. My opinion is that the basic ideas and techniques used are already so difficult for the uninitiated that further definitions, axioms, and abstractions would tend to obfuscate rather than clarify. (Such abstractions are occasionally pointed to, however.)

I gratefully acknowledge the support and encouragement of the members of the Subcommittee on Carus Monographs during the preparation of this volume: Daniel Finkbeiner, George Piranian, and Ralph Boas. Both Daniel Finkbeiner and Wendell Lindstrom labored through the entire text and made numerous valuable suggestions. John Fay critically read parts of the manuscript, and Victor Albis, with mathematical and bibliographical assistance, rescued me from despair during the writing of the last chapter. Support received during a year at Bowdoin College, especially the availability of the outstanding Smyth Mathematical Library, contributed significantly to this project. I apologize for any errors that may be found.

I now take more seriously other authors' expressions of gratitude to spouses and children for their sacrifices and long suffering during such writing projects. To my wife Joanne and to my children, I express my thanks for such support and such endurance. If there will be readers who through this book find the same excitement and enjoyment in the material as I did, then I think it will all have been worth it.

CHARLES R. HADLOCK

Cambridge, Massachusetts

CONTENTS

INTRODUCTION

Mathematical knowledge has undergone explosive growth in the past few hundred years. No doubt, a significant portion of this growth is due to the usefulness of mathematics as a tool in many fields. However, probably the greatest single stimulus to the development of this subject is its natural intellectual appeal—the challenge of a hard problem and the satisfaction of resolving it. For centuries these factors have motivated the efforts of mathematicians.

Problems which are simple to state and understand are quite reasonably often the most attractive. It is not uncommon for such problems to stimulate extensive research by many mathematicians, and through their solution to lead to results and insights far beyond their original bounds. The solution of one problem inevitably leads to the posing of others, and thus the subject of mathematics continues to grow.

The theories emerging from the solution of fundamental problems continue to be refined and generalized, for part of the work of the mathematician is to abstract from one situation its fundamental features and to apply these to other situations that are similar. A mature theory may bear scant resemblance to the solution of the original problems upon which it is founded. And the abstract exposition of such a theory may not stimulate in those who would study it the intellectual excitement and curiosity which led to its development.

What are some of these early problems that have given birth to great theories? In this book, we shall be concerned with the following:

A. The Three Greek Problems

Most people have learned in school how to perform certain geometric constructions by means of a compass and an unmarked straightedge. For example, it is quite simple to bisect an angle, to construct the perpendicular bisector of a line segment, or to construct the line through a given point parallel to a given line. With the same instruments and rules of construction, is it possible to

(1) *double a cube*, meaning to construct the edge of a cube whose volume would be twice that of a cube with a given segment as an edge;

(2) *trisect an angle*, meaning to divide an arbitrary angle into three equal parts; or

(3) *square a circle*, meaning to construct a square having the same area as a given circle?

These problems date from the fifth century B.C., and they have a long and interesting history. For example, legend has it that ancient Athens, being faced by a serious plague, sent a delegation to the oracle of Apollo at Delos for advice in their difficulty. The delegation was told to double the cubical altar to Apollo. Unfortunately, they doubled the length of each edge, thereby increasing the volume by a factor of eight rather than two; and the plague only got worse. From this story, the problem of doubling a cube has received the name "the Delian problem". For detailed information on the history of these problems, the reader may wish to consult a book on the history of mathematics, several of which are listed later in this section.

With respect to Problems 1 and 3, the construction is never possible. With respect to Problem 2, there exist angles that cannot be trisected, among them the angle of 60°. One should pay careful attention to the nature of these results. They do not simply assert that no construction is known, that none has been discovered; they actually say that *none can exist*, which is a much stronger statement. It is natural to wonder how such a negative result might be established. It will not be necessary to wait long, as these problems are solved in the first chapter.

Historically, however, it took about 2200 years before solutions were discovered, despite frequent periods of frantic effort. It is hard to say just when Problem 1 was first solved. A solution to Problem 2 came as a special case of more general results by Carl Friedrich Gauss (1777–1855) and Pierre Wantzel (1814–1848) on the problem of finding those values of n for which it is possible to construct a regular polygon with n sides. Gauss made substantial progress on this problem by the age of eighteen! (We shall take up this latter problem in Chapter 2.) Problem 3 was solved by F. Lindemann in 1882 [Über die Zahl π, *Math. Ann.*, 20 (1882), 213–225]; it is somewhat more difficult than the first two. For a fascinating history of many false solutions to this problem (including a variety of methods—mathematical, religious, etc.), it is well worth searching out a copy of the book by

F. J. Duarte, *Monografía sobre los Números π y e* [Estados Unidos de Venezuela, *Bol. Acad. Ci. Fís. Mat. Nat.*, 11 (1948), no. 34–35, 1–252 (1949); *Math. Reviews*, 11–501].

B. The Solution of Polynomial Equations by Radicals

The reader is probably familiar with the 'quadratic formula', which asserts that the solutions to the quadratic equation

$$ax^2 + bx + c = 0$$

are given by

$$x = \frac{-b \pm \sqrt{b^2 - 4ac}}{2a}.$$

Can a similar formula be found for the general polynomial equation of degree 3:

$$ax^3 + bx^2 + cx + d = 0\ ?$$

How about polynomials of degree 4, 5, or, in general, n? By a 'similar formula', let us agree to mean some definite procedure for calculating the solutions from the coefficients of the polynomials, using the rational operations (addition, subtraction, multiplication, and division) and the extraction of roots. Solutions obtained using these operations are called 'solutions by radicals'.

It turns out that for polynomials of degree less than or equal to 4, such a procedure exists, whereas for polynomials of higher degree, no such general procedure exists. Indeed, there are polynomials of every degree greater than or equal to 5 with roots which cannot be obtained from the coefficients by any sequence of rational operations and the extraction of roots. Again, this is a very strong negative result.

The formula for the quadratic equation seems first to have been discovered by the Moslems around 900 A.D., although quadratic equations had been solved by the Babylonians 3000 years earlier. The general cubic equation was probably first solved by Scipione del Ferro (1465–1526) and the quartic equation by Ludovico Ferrari (1522–1565). For the next few centuries, considerable efforts were directed towards the quintic equation, with the expectation that a general formula would eventually be found. However, towards the end of the eighteenth century, a number of mathematicians began to suspect that perhaps it was not possible to obtain a formula for the solution of the general quintic equation. The work of Joseph Lagrange (1736–1813) had considerable effect on this change of attitude. Beginning in 1799, Paolo Ruffini (1765–1833) published several somewhat deficient proofs of this negative result. Finally, Niels Henrik Abel (1802–1829) produced in 1828 an essentially correct proof of the unsolvability of the general quintic. Soon thereafter, at the age of nineteen, Évariste Galois (1811–1832) perfected the theory by giving a simple necessary and sufficient condition for the solvability by radicals of a polynomial equation. The theorems of Abel and Galois are developed in Chapter 3 of this book.

While it is far from apparent in their statement, the solutions of Problems A and B use many of the same ideas and techniques. These ideas have led to the modern theories of groups and fields. Rather than focus on the abstract theories in this book, we shall simply imagine at the outset that we face these problems, and as we proceed we shall develop the ideas necessary for their solution. It is hoped that the concreteness of this approach will serve to invite the readers to ask their own questions, to wonder freely, and to pursue matters of individual interest along the way. At the end, a powerful theory will have been developed, and to convert it to its usual abstract formulation is then a relatively small and easy step. For this abstract formulation, the reader may wish to consult textbooks on abstract algebra.

The lines of the development here are not, in general, those of the original proofs; rather, they are heavily influenced by modern approaches and terminology. The reader may be interested to take a look at some of the original papers on this material, especially the works of Abel, Galois, and Gauss. The references are:

Niels Henrik Abel, *Oeuvres complètes*, 2 vols., ed. by L. Sylow and S. Lie, Grondahl and Sons, Christiana, 1881.

Évariste Galois, *Écrits et Mémoires Mathématiques*, ed. by R. Bourgne and J.-P. Azra, Gauthier-Villars, Paris, 1962.

Carl Friedrich Gauss, *Werke*, 9 vols., Königliche Gesellschaft der Wissenschaften, Göttingen, 1863–1906.

———, *Disquisitiones Arithmeticae*, tr. by A. Clarke, Yale University Press, New Haven, 1966.

While it is easy to pass such references by, even a cursory examination of the original works lends a certain valuable historical perspective to the mathematical developments. (And the scrap papers and doodlings of these famous mathematicians are sometimes the greatest contributors in this respect. There are some samples of Galois' in his collected works; for Abel, photostats are available in the biography by Oysten Ore [Univ. of Minnesota Press, 1957].) An extensive paper tracing the historical evolution of the theory into its modern form has been written by B.

Melvin Kiernan [The Development of Galois Theory from Lagrange to Artin, *Arch. History Exact Sci.*, 8 (1971–72), 40–154]. It may also be interesting to read concurrently some material on the history of mathematics. Good sources are:

E. T. Bell, *Men of Mathematics*, Simon and Schuster, New York, 1937.

Carl Boyer, *A History of Mathematics*, Wiley, New York, 1968.

Morris Kline, *Mathematical Thought from Ancient to Modern Times*, Oxford University Press, New York, 1972.

David Eugene Smith, *History of Mathematics*, vols. 1 and 2, Dover, New York, 1958.

One cannot help but be awed by the magnificent accomplishments of Abel and Galois, especially during their short and tragic lives, the former having died at age twenty-nine and the latter having been killed in a foolish duel before the age of twenty-one!

Before ending the Introduction, it is necessary to say a word about the problem sets at the end of each section in the main text. Mathematics is not a "spectator sport"; it is necessary to read mathematics with paper and pencil at hand. The problems at the end of each section are extremely important. One should read through all the problems, for many of them are referred to later on. A few minutes should be spent on each, enough to think through the solution in cases when the problem is very simple, or to plan a possible attack in more difficult cases. Then, those problems that are appealing to the reader should be tried in their entirety. Particularly challenging problems are marked with an asterisk. Lastly, for any problems that the reader has thought about but has not been inclined or able to give a complete solution, it may be helpful to consult the solutions given in the back of the book. But—and this is important—it is largely unproductive to read solutions before at least some thought has been given to the problems.

It should be kept in mind that the persons who first solved the problems in this book probably did so largely because of the sheer intellectual appeal and challenge of the problems; they were fun. It is hoped that the reader will find similar attractions in this material.

CHAPTER 1

THE THREE GREEK PROBLEMS

Section 1.1. Constructible Lengths

An important step in the solution of the classical Greek construction problems will be taken in this section, where we shall answer the question: **Given a line segment of length 1 in the plane, for what values of a can we construct a segment of length a?** As usual in geometry, we shall always consider lengths to be non-negative. To begin, naturally, we need to take a careful look at the rules which govern the construction, where by "construction" we refer to a finite sequence of operations with a compass and an unmarked straightedge. The individual operations which may be done with these instruments are called the *fundamental constructions*; they are:

(1) Given 2 points, we may draw a line through them, extending it indefinitely in each direction.

(2) Given 2 points, we may draw the line segment connecting them.

(3) Given a point and a line segment, we may draw a circle with center at the point and radius equal to the length of the line segment.

It is tempting to add one additional operation, namely:

Given a point, we may draw a circle with center at that

A circle (or an arc thereof) is drawn with center P and any radius greater than the perpendicular distance from P to the line.

Figure 1

point and indefinite radius, possibly restricted by some inequality.

The reader may recall such an operation, for example, in the first step in constructing the line perpendicular to a given line and through a given point P not on that line, as in Figure 1. However, it will follow easily from Lemma 1a that, given a segment of unit length, all segments of rational lengths are constructible. Hence, within any restricted interval of lengths, we can always find one of the type prescribed in operation 3. Thus it is without loss of generality that the fourth operation is omitted.

Using the simple constructions of high school geometry, it is possible to show:

LEMMA 1a. *Given segments of lengths* 1, a, *and* b, *it is possible to construct segments of lengths* $a+b$, $a-b$ (*when* $a>b$), ab, *and* a/b (*when* $b \neq 0$).

Proof. The first two should be obvious. To obtain the product ab, first construct two non-collinear rays emanating from a single point; for example, construct two perpendicular rays. (You may follow this construction on the first triangle in Figure 2.) It should be clear that this can be accomplished beginning only with a single (unit) segment. On these rays, mark off segments of lengths a and 1 respectively, and connect the resulting points to form a triangle. On the ray on which the length 1 has been marked off, now mark off a length b also from the end. If a line is now constructed through this point parallel to the third side of the triangle constructed earlier, it intersects the other ray at a distance ab from the endpoint of the ray, as follows at once by the similarity of the triangles. The second triangle in Figure 2 pertains to the analogous argument for the quotient a/b, for which the reader should think through the steps. ∎

Let us agree to call a real number, a, *constructible* if, given initially a segment of length 1, it is possible to construct a segment of length $|a|$. By consideration of

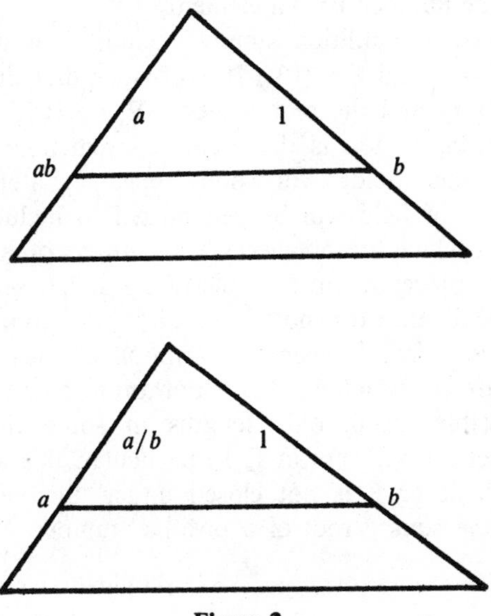

Figure 2

various cases for the signs and relative sizes of numbers a and b, it is an immediate consequence of Lemma 1a that if a and b are constructible, then so too are $a+b$, $a-b$, ab, and, when $b \neq 0$, a/b. (Cf. Problem 1.) From this it follows that all the rational numbers are constructible, for any integer can be formed by adding or subtracting an appropriate number of 1's, and so then any quotient of integers can be obtained.

We shall find it very convenient in our work to study certain special sets of numbers, called *fields*. Let **F** be a subset of the set **R** of real numbers. We say that **F** is a *field* if it satisfies the following two conditions:

(1) **F** is *closed* under the *rational operations* (addition, subtraction, multiplication, and division except by 0), meaning that whenever these operations are applied to elements of **F**, the result is an element of **F**.

(2) The number 1 is an element of **F**.

(The second condition simply excludes the two trivial cases $\mathbf{F} = \emptyset$ and $\mathbf{F} = \{0\}$.) It is obvious that the rational numbers **Q** and the real numbers **R** are fields; and, by Lemma 1a, so too is the set of constructible numbers. Many other fields will soon appear. (Later on, the definition of field will be generalized to include sets of numbers which are not necessarily subsets of **R**.) The set **Z** of all integers is an example of a familiar set which is not a field, since it is not closed under division.

In our search for some description of those numbers which are constructible, it is important to note that it may be possible, using only lengths in some field **F**, to construct a length not in **F**. In particular, this will be the case whenever **F** is not closed under the operation of taking the square root of a positive number, as we now show.

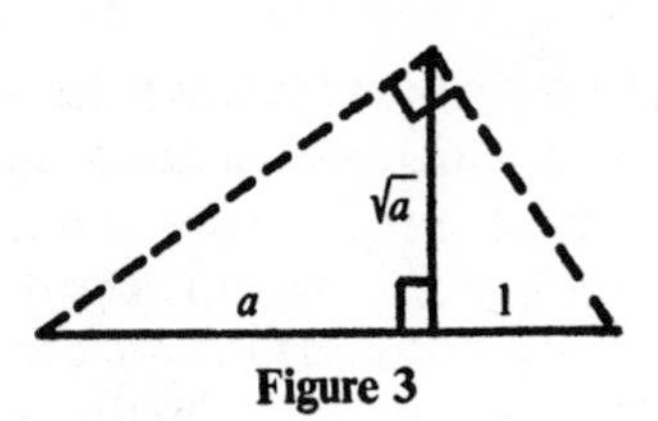

Figure 3

LEMMA 1b. *Given segments of lengths* 1 *and* a, *a segment of length* $\sqrt{a}$ *may be constructed.*

Proof. Figure 3 indicates one such construction. Adjacent segments of lengths a and 1 are marked off on a line. The midpoint of this combined segment is then constructed, and from this the circle having the segment for a diameter. A perpendicular constructed from the point where the original two segments were joined intersects the circle at a distance $\sqrt{a}$ from the diameter, as follows from the similarity of the two small triangles in the diagram. ∎

Thus we are able to deduce from this and the previous lemma that if all the numbers of a field **F** are constructible, then so too are all the numbers of the form $a + b\sqrt{k}$, where a, b, and k are in **F** and $k > 0$. What is interesting is that for a fixed element $k \in \mathbf{F}$, $k > 0$, the set of all numbers of the form $a + b\sqrt{k}$ itself forms a field! It is called the *extension of* **F** *by* $\sqrt{k}$ and is denoted $\mathbf{F}(\sqrt{k})$. Of course, if $\sqrt{k}$ is in **F**, this set is simply **F** itself. But if $\sqrt{k}$ is not in **F**, this new set contains **F** as a proper subset. In this case $\mathbf{F}(\sqrt{k})$ is called a *quadratic extension* of **F**.

LEMMA 1c. *If* **F** *is a field and if* $k \in \mathbf{F}$, $k > 0$, *then* $\mathbf{F}(\sqrt{k})$ *is also a field.*

Proof. Since $1 \in \mathbf{F}$, $1 \in \mathbf{F}(\sqrt{k})$. All we need to show now is that the sum, difference, product, and quotient of numbers in $\mathbf{F}(\sqrt{k})$ are themselves in $\mathbf{F}(\sqrt{k})$, that is, that they have the form $A + B\sqrt{k}$ with $A, B \in \mathbf{F}$. We have:

$$(a + b\sqrt{k}) \pm (c + d\sqrt{k}) = (a \pm c) + (b \pm d)\sqrt{k},$$

$$(a + b\sqrt{k})(c + d\sqrt{k}) = (ac + bdk) + (ad + bc)\sqrt{k},$$

$$\frac{a + b\sqrt{k}}{c + d\sqrt{k}} = \frac{a + b\sqrt{k}}{c + d\sqrt{k}} \cdot \frac{c - d\sqrt{k}}{c - d\sqrt{k}}$$

$$= \left(\frac{ac - bdk}{c^2 - d^2k}\right) + \left(\frac{bc - ad}{c^2 - d^2k}\right)\sqrt{k}.$$

Of course, with respect to the division part, we must be careful that our denominators are not zero. By assumption, $c + d\sqrt{k} \neq 0$. If it happens to be the case that $c - d\sqrt{k} = 0$, it follows that $c + d\sqrt{k} - 0 = c + d\sqrt{k} - (c - d\sqrt{k}) = 2d\sqrt{k} \neq 0$, and so $d \neq 0$. But then we would have $\sqrt{k} = c/d \in \mathbf{F}$, and so the original quotient would be in $\mathbf{F}$ and hence $\mathbf{F}(\sqrt{k})$ by inspection. In any case, we see that the result of each of the arithmetic operations is a number of the required form. ∎

We know now that every rational number is constructible and that every number that can be calculated from the rationals by a finite sequence of the rational operations and the square root operation is also constructible. For example, the number $\sqrt[4]{13} + \frac{4}{3}\sqrt{\sqrt{6} + \sqrt{1 + 2\sqrt{7}}}$ is constructible, as we can build it up by the following sequence:

$$7, \sqrt{7}, \sqrt{7}+\sqrt{7}=2\sqrt{7}, 1, 1+2\sqrt{7}, \sqrt{1+2\sqrt{7}},$$

$$6, \sqrt{6}, \sqrt{6}+\sqrt{1+2\sqrt{7}}, \sqrt{\sqrt{6}+\sqrt{1+2\sqrt{7}}}, \tfrac{4}{3},$$

$$\tfrac{4}{3}\sqrt{\sqrt{6}+\sqrt{1+2\sqrt{7}}}, 13, \sqrt{13}, \sqrt{\sqrt{13}}=\sqrt[4]{13},$$

$$\sqrt[4]{13}+\tfrac{4}{3}\sqrt{\sqrt{6}+\sqrt{1+2\sqrt{7}}}.$$

This operation may be conveniently restated in terms of fields:

LEMMA 1d. *A number a is constructible if there is a finite sequence of fields $\mathbf{Q}=\mathbf{F}_0 \subset \mathbf{F}_1 \subset \cdots \subset \mathbf{F}_N$, with $a \in \mathbf{F}_N$, and such that for each j, $0 \leqslant j \leqslant N-1$, $\mathbf{F}_{j+1}$ is a quadratic extension of $\mathbf{F}_j$.*

Proof. By induction on N. If $N=0$, a is rational and hence constructible. Now we assume the theorem is true for $N=R$ and we prove it for $N=R+1$. If $a \in \mathbf{F}_{R+1}$, then a may be expressed as $a_R + b_R\sqrt{k_R}$, where a_R, b_R, and k_R all belong to $\mathbf{F}_R$. By the inductive hypothesis, a_R, b_R, and k_R are all constructible; and so, by Lemmas 1b and 1a, $a = a_R + b_R\sqrt{k_R}$ is constructible. ∎

As an example, one such sequence of fields for the number $\sqrt[4]{13}+\frac{4}{3}\sqrt{\sqrt{6}+\sqrt{1+2\sqrt{7}}}$, which was treated just before the lemma, is:

$$\mathbf{F}_0 = \mathbf{Q}, \mathbf{F}_1 = \mathbf{F}_0(\sqrt{7}), \mathbf{F}_2 = \mathbf{F}_1\left(\sqrt{1 + 2\sqrt{7}}\right),$$

$$\mathbf{F}_3 = \mathbf{F}_2(\sqrt{6}), \mathbf{F}_4 = \mathbf{F}_3\left(\sqrt{\sqrt{6} + \sqrt{1 + 2\sqrt{7}}}\right),$$

$$\mathbf{F}_5 = \mathbf{F}_4(\sqrt{13}), \mathbf{F}_6 = \mathbf{F}_5\left(\sqrt{\sqrt{13}}\right),$$

which is a field containing the number. Such a sequence is simply determined by following through the steps of the construction and extending the field each time a square root not already in the field is introduced. If one were actually interested in proving that one had a sequence of fields precisely as described in the lemma, namely a sequence of quadratic extensions, it might take some work to determine at which steps a truly larger field is produced. For example, could you show in the case above that $\sqrt{13} \notin \mathbf{F}_4$? For our purposes, such calculations will not be important. Incidentally, there are many different paths that one could follow in building up the complicated number above, or any others. Different paths may give different sequences of fields. (See Problem 9.)

Lemma 1d characterizes a certain class of constructible numbers. **Are there any other constructible numbers?** The answer is "no"; in order to see this, we shall make use of some elementary analytic geometry.

If **F** is a field, the *plane of* **F** will denote the set of all points (x,y) in the Cartesian plane such that both x and y are in **F**. By a *line in* **F** is meant a line passing through two points in the plane of **F**. By a *circle in* **F** is meant a circle with both its center and some point on its circumference in the plane of **F**. The importance of these notions is the fact that any fundamental construction using only points in the plane of a field **F** involves the construction of (a portion of) a line or a circle *in* **F**. For the first two fundamental constructions this is obvious by the definition of a line in **F**. For the third fundamental construction, that of a circle with a given center and a radius equal to the length of the segment connecting two given points, suppose that the center is (x_1, y_1) and the segment is that connecting (x_2, y_2) and (x_3, y_3), where all the coordinates are in **F**, as in Figure 4. It is now easy to see that the desired circle is in the plane of **F**, for its center is in the plane of **F** and it passes through $(x_1 + x_3 - x_2, y_1 + y_3 - y_2)$, which is also in the plane of **F**.

We find further that lines and circles in **F** can be described by simple equations, all of whose coefficients are themselves in **F**, as stated below:

LEMMA 1e. *Every line in* **F** *can be represented by an equation of the form* $ax + by + c = 0$, *where* a, b, *and* c *are all in* **F**. *Every circle in* **F** *can be represented by an equation of the form* $x^2 + y^2 + ax + by + c = 0$, *where* a, b, *and* c *are all in* **F**.

Proof. In the case of the line, let (x_1,y_1) and (x_2,y_2) be two points on it and in the plane of **F**. If $x_1 = x_2$, then of course the line is described by the equation $x = x_1$, which has the desired form ($a = 1$, $b = 0$, $c = -x_1$). If $x_1 \neq x_2$, then the equality of slopes,

$$\frac{y - y_1}{x - x_1} = \frac{y_2 - y_1}{x_2 - x_1},$$

yields the equation

$$(y_2 - y_1)x + (x_1 - x_2)y + (y_1x_2 - x_1y_2) = 0,$$

which is also of the desired form.

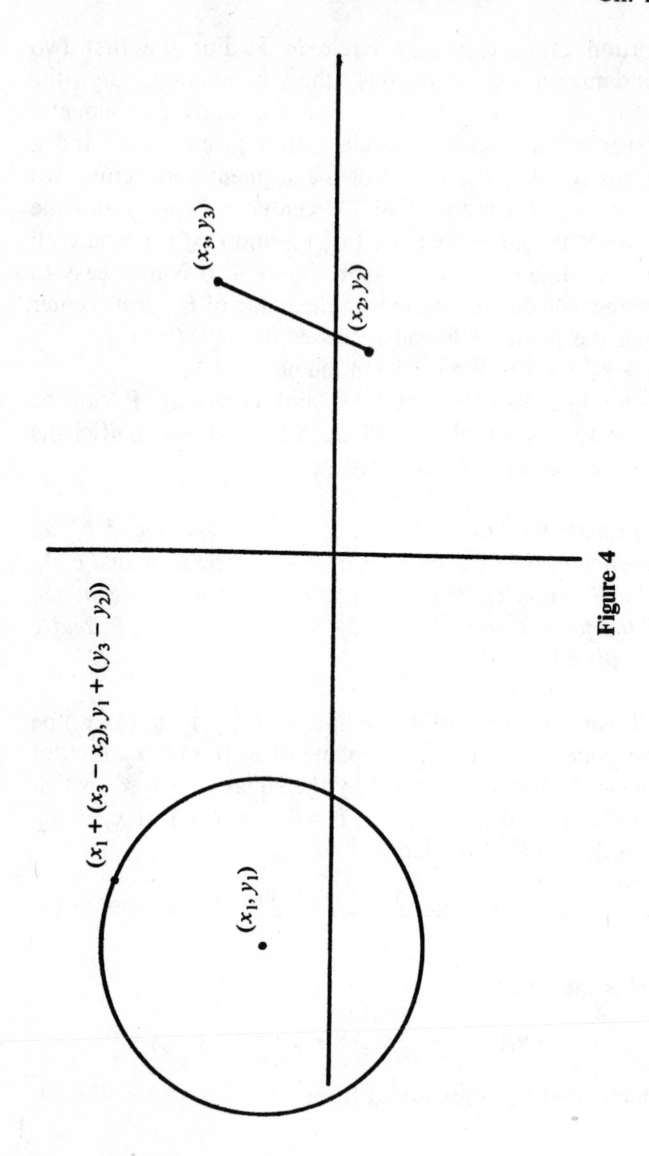

Figure 4

In the case of the circle, denoting the center by (x_1,y_1) and a point on it and in the plane of **F** by (x_2,y_2), the equation is

$$(x-x_1)^2+(y-y_1)^2=(x_2-x_1)^2+(y_2-y_1)^2.$$

This clearly reduces to the desired form. ∎

Since the only way to construct a segment is by constructing its endpoints, and since the only way to construct a point is as the intersection of two lines, a line and a circle, or two circles, it is important to determine the nature of such intersections. In particular, we have:

LEMMA 1f. *The point of intersection of two lines in* **F** *is in the plane of* **F**. *The points of intersection of a line in* **F** *and a circle in* **F**, *as well as the points of intersection of two circles in* **F**, *are either in the plane of* **F** *or in the plane of some quadratic extension of* **F**.

(Note that the plane of **F** is contained in the plane of any $\mathbf{F}(\sqrt{k})$, so that the words "either in the plane of **F** or" could be omitted without changing the meaning of the lemma. It is more convenient to think of it in the form given, however.)

Proof. The case of the intersection of two lines in **F** amounts to the simultaneous solution of the two equations

$$a_1x+b_1y+c=0$$

$$a_2x+b_2y+c=0$$

where the coefficients are, by Lemma 1e, all in **F**. It is obvious that this only involves rational operations, and so the solutions x and y are in **F**, which means that (x,y) is in the plane of **F**.

The case of a line in **F** and a circle in **F** amounts to the simultaneous solution of the two equations

$$a_1x + b_1y + c_1 = 0$$

$$x^2 + y^2 + a_2x + b_2y + c_2 = 0$$

where the coefficients are all in **F**. Since a_1 and b_1 cannot both be 0, the first equation can be used to solve for one of the variables in terms of the other. Without loss of generality we assume that we can solve for y:

$$y = \frac{-c_1}{b_1} - \frac{a_1}{b_1}x.$$

When this is substituted into the second equation, we obtain a quadratic equation in x with coefficients in **F**. From the quadratic formula, its solution, which we need not calculate explicitly, has the form $A \pm B\sqrt{k}$, with A, B, and k in **F** and $k \geqslant 0$ (otherwise the line doesn't intersect the circle). Substituting this in the equation for y also yields an expression of the form $A' \pm B'\sqrt{k}$. Thus the resulting points are either both in the plane of **F**, as when $\sqrt{k} \in \mathbf{F}$, or both in the plane of $\mathbf{F}(\sqrt{k})$, $\sqrt{k} \notin \mathbf{F}$. (When $k = 0$, of course, we only have one point; and it is in the plane of **F**.)

For the case of two circles, the equations

$$x^2 + y^2 + a_1x + b_1y + c_1 = 0$$

$$x^2 + y^2 + a_2x + b_2y + c_2 = 0$$

can be subtracted, yielding a linear equation, with coefficients in **F**, to be solved simultaneously with the equation of either of the circles. Thus this reduces to the situation in the previous case. ∎

We have finally reached the point where it is possible to state and prove the main result concerning constructibility, thus answering the question posed in the first sentence of this section.

THEOREM 1. *The following two statements are equivalent:*

(i) *The number a is constructible.*

(ii) *There exists a finite sequence of fields $\mathbf{Q} = \mathbf{F}_0 \subset \mathbf{F}_1 \subset \cdots \subset \mathbf{F}_N$, with $a \in \mathbf{F}_N$, and such that for each j, $0 \leqslant j \leqslant N-1$, $\mathbf{F}_{j+1}$ is a quadratic extension of $\mathbf{F}_j$.*

Proof. (ii)$\Rightarrow$(i). This is simply a restatement of Lemma 1d.

(i)$\Rightarrow$(ii). If we are given the points $(0, 0)$ and $(1, 0)$ in the Cartesian plane, then by hypothesis a segment of length $|a|$ can be constructed. It is clear that we may use this segment, the point $(0, 0)$, and the line through $(0, 0)$ and $(1, 0)$, to construct the point $(a, 0)$, which we shall call P. Thus it suffices to show that P is in the plane of a field $\mathbf{F}_N$ of the type described in (ii), that is, a field obtainable from $\mathbf{Q}$ by a sequence of quadratic extensions.

Now, the construction of P involves a finite number of the fundamental constructions, each of which results in a finite number of new points as intersections of various types. We list all such points in order of construction, where points resulting on the same step may be listed in any order. The point P is on this list, say in the Mth position. Omitting anything in the list after P (which could only include points constructed on the same step as P), we have: $P_1, P_2, \ldots, P_{M-1}, P_M$. The theorem will be proved once we have established the following claim: There exists a field $\mathbf{F}$, obtainable from $\mathbf{Q}$ by a sequence of quadratic extensions, such that $P_1, P_2, \ldots, P_M$ are all in the plane of $\mathbf{F}$. Since P_1 and P_2 must be the two given points $(0, 0)$ and $(1, 0)$, which are in the plane of $\mathbf{Q}$, the claim holds for $M = 1$ and 2. To show that it holds for any M, we recall that the construction of P_M only involves figures constructed using (some subset of) the points P_1 through P_{M-1}, and hence, by the inductive hypothesis, figures in the plane of some field $\tilde{\mathbf{F}}$ obtainable from $\mathbf{Q}$ by a sequence of quadratic extensions. But then, by Lemma 1f, P_M is in the plane of either $\tilde{\mathbf{F}}$ or $\tilde{\mathbf{F}}(\sqrt{k}\,)$, for some $k \in \tilde{\mathbf{F}}$ with $\sqrt{k} \notin \tilde{\mathbf{F}}$. In either case, P_M, as well as the previous P_i's, are all in the plane of a field of the required type. ∎

The reader should take note of the fact that this theorem sets up an equivalence between two properties, one of which is geometric and one of which is algebraic. It is the conversion of our classical geometric problems to algebraic problems, via this theorem, that provides the key to their solution. This will be shown in the next few sections.

PROBLEMS

1. Use Lemma 1a to show that if a and b are constructible, then so too are $a + b$, $a - b$, ab, and, when $b \neq 0$, a/b.

2. If M and N are positive integers, show that if $\sqrt[M]{N}$ is rational, then, in fact, it must be an integer. Conclude then that $\sqrt{2}$, $\sqrt{3}$, $\sqrt[3]{2}$, $\sqrt{6}$, for example, are all

irrational. (Hint: If $\sqrt[M]{N}$ is rational it may be expressed in lowest terms as Q/R. If $R \neq \pm 1$ it has a prime factor. If a prime divides a product of integers, it must divide one of them.)

*3. We know from Lemma 1c that if $\mathbf{F}$ is a field, so too is the set $\{a+b\sqrt{2} \mid a, b \in \mathbf{F}\}$. What about the set $\{a+b\sqrt[3]{2} \mid a, b \in \mathbf{F}\}$? Is it ever a field? Is it always a field? Give as complete an analysis as you can.

4. Let $\mathfrak{F}$ be a collection of fields. Show that the common intersection of all the fields in $\mathfrak{F}$, written $\bigcap_{\mathbf{F} \in \mathfrak{F}} \mathbf{F}$, is itself a field.

*5. Give an explicit description of the *smallest field* containing $\sqrt[3]{2}$. (By the *smallest* field having a given property is meant the intersection of all fields with that property. Problem 4 guarantees that such an intersection is a field.)

*6. If p is a prime and N is a positive integer $\geqslant 2$, give an explicit description of the smallest field containing $\sqrt[N]{p}$. (This is a generalization of the previous problem.)

7. Is $\{a+b\sqrt{2}+c\sqrt{3} \mid a, b, c \in \mathbf{Q}\}$ a field?

8. Give an explicit representation of the smallest field containing both $\sqrt{2}$ and $\sqrt{3}$. (Hint: Use one of the lemmas.)

9. For the example treated after Lemma 1d, give a different development of the same number from the rationals which results in a different sequence of field extensions.

10. Let $\mathbf{F}$ be a field. Is the set $\{a+b\sqrt{k} \mid a, b, k \in \mathbf{F}, k>0\}$ necessarily a field? (This differs from $\mathbf{F}(\sqrt{k})$ because in the present case k is not fixed, but rather it ranges through all positive values in $\mathbf{F}$.)

11. The following argument is occasionally found in

proofs of Theorem 1: "Using only points in the plane of a field **F**, by a single fundamental construction it is only possible to construct points in the plane of some $\mathbf{F}(\sqrt{k})$, $k \in \mathbf{F}$." Criticize this argument or use it to give a simpler proof of Theorem 1.

12. What else would you like to know? Spend a little time thinking about the material in this section and perhaps especially in the problems above. Can you anticipate the solution of any of the classical construction problems: doubling the cube, trisecting an angle, squaring the circle? Are you curious about anything? Do any questions occur to you that seem interesting? Make a short list of your questions and ideas, spending at least enough time on each to satisfy yourself that it is nontrivial. Keep these ideas in mind, jotting down others as they occur to you as you proceed through the material in subsequent sections.

Section 1.2. Doubling the Cube

We begin by recalling the classical problem of "doubling the cube": **Given a line segment representing the edge of a cube, is it possible to construct another line segment representing the edge of a cube of twice the volume of the first?** If we take our unit of length to be the length of the given segment, then the desired segment must have length $\sqrt[3]{2}$. So the problem is simply to determine whether $\sqrt[3]{2}$ is constructible. By Theorem 1, if $\sqrt[3]{2}$ were to be constructible, there would have to be associated with it a sequence of quadratic field extensions. That this is impossible will follow easily from the following lemma, which also happens to give a complete answer to the question raised in Problem 3 of the previous section.

LEMMA 2. *Let* $\mathbf{F}(\sqrt{k})$ *be a quadratic extension of a field* $\mathbf{F}$. *If* $\sqrt[3]{2}$ *is in* $\mathbf{F}(\sqrt{k})$, *then* $\sqrt[3]{2}$ *must be in* $\mathbf{F}$ *itself.*

Proof. We are given that $\sqrt[3]{2}$ may be written in the form $a + b\sqrt{k}$, with $a, b, k \in \mathbf{F}$, $\sqrt{k} \notin \mathbf{F}$. We want to show that b must be 0. The calculation

$$\begin{aligned}
2 &= (a + b\sqrt{k})^3 \\
&= a^3 + 3a^2b\sqrt{k} + 3ab^2k + b^3k\sqrt{k} \\
&= [a^3 + 3ab^2k] + [3a^2b + b^3k]\sqrt{k}
\end{aligned}$$

implies that $3a^2b + b^3k = 0$, for otherwise we could solve for $\sqrt{k}$ which would then have to be an element of $\mathbf{F}$. But then we also have

$$(a - b\sqrt{k})^3 = [a^3 + 3ab^2k] - [3a^2b + b^3k]\sqrt{k} = 2,$$

which gives us another cube root of 2, namely $a - b\sqrt{k}$. Since the function $y = x^3$ is strictly increasing, there is at most one real cube root of 2. This requires that $b = 0$. ∎

And now the main result follows:

THEOREM 2. *It is impossible to "double the cube".*

Proof. As noted in the introductory remarks, doubling the cube is equivalent to constructing $\sqrt[3]{2}$. If $\sqrt[3]{2}$ were constructible, by Theorem 1 there would exist a sequence of fields $\mathbf{Q} = \mathbf{F}_0 \subset \mathbf{F}_1 \subset \mathbf{F}_2 \subset \cdots \subset \mathbf{F}_N$, each a quadratic extension of the previous one, such that $\sqrt[3]{2} \in \mathbf{F}_N$. The repeated application of Lemma 2 obviously implies that $\sqrt[3]{2} \in \mathbf{Q}$, which contradicts the irrationality of $\sqrt[3]{2}$. (The irrationality of $\sqrt[3]{2}$ may be established very simply as in Problem 2 of the previous section.) ∎

The basic idea of this section will be slightly generalized in the next section, where it will yield the solution of the angle trisection problem.

PROBLEMS

1. Can the cube be "tripled"?
2. Can the cube be "quadrupled"?

Section 1.3 Trisecting the Angle

The classical problem involving angle trisection is: **Given an arbitrary angle, is it possible to trisect it?** (As usual, we are restricting ourselves to the use of a compass and an unmarked straightedge.) The answer is "no", but one must be careful to understand what this answer means. There are certainly some particular angles that can be trisected, as, for example, the angle of 90°. (Why?) However, it is not possible to trisect an *arbitrary* angle. To put it another way, there exists at least one angle that *cannot* be trisected. We shall actually produce one such angle, in particular the innocuous-looking angle of 60°.

Since a 60° angle can be constructed, if it could also be trisected, then a 20° angle could be constructed. But then, as illustrated in Figure 5, the number cos 20° would be constructible. It will be shown, however, that the number cos 20° is a root of a cubic equation which has no constructible roots, thus leading to a contradiction.

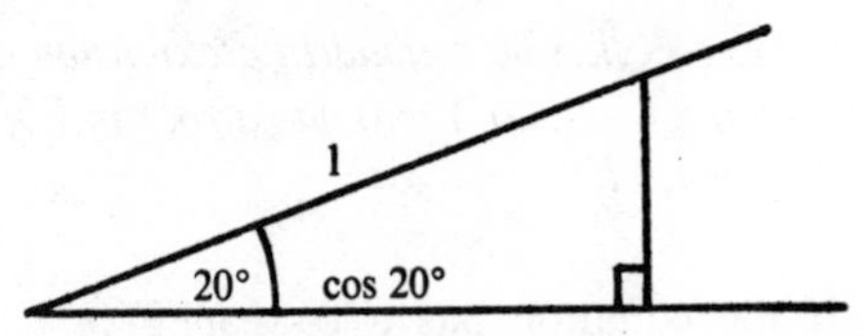

Mark off a unit length on one side and drop a perpendicular to the other.

Figure 5

From elementary trigonometric formulas, we have

$$\begin{aligned}\cos 3\theta &= \cos(2\theta + \theta)\\ &= \cos 2\theta \cos \theta - \sin 2\theta \sin \theta\\ &= (\cos^2\theta - \sin^2\theta)\cos \theta - (2 \sin \theta \cos \theta)\sin \theta\\ &= \cos^3\theta - 3 \sin^2\theta \cos \theta\\ &= \cos^3\theta - 3(1 - \cos^2\theta)\cos \theta\\ &= 4 \cos^3\theta - 3 \cos \theta.\end{aligned}$$

When $\theta = 20°$, $\cos 3\theta = \frac{1}{2}$, so the number $\cos 20°$ must be a root of the cubic equation

$$8u^3 - 6u - 1 = 0.$$

This equation is made a little simpler by the change of variable $x = 2u$, which yields

$$x^3 - 3x - 1 = 0.$$

If $\cos 20°$ were constructible, then it would be possible to construct a root of this equation. As the first step in the proof that this is impossible, we make an observation which is very similar to that of Lemma 2:

LEMMA 3. *Let* $\mathbf{F}(\sqrt{k})$ *be a quadratic extension of a field* $\mathbf{F}$. *If the equation* $x^3 - 3x - 1 = 0$ *has a root in* $\mathbf{F}(\sqrt{k})$, *then it has a root in* $\mathbf{F}$ *itself.*

Proof. If the equation has a root in $\mathbf{F}(\sqrt{k})$, we may denote it by $a + b\sqrt{k}$, where a, b, and k are in $\mathbf{F}$. If $b = 0$ we are done, for $a \in \mathbf{F}$ is a root. If $b \neq 0$, it will be shown that $-2a$ is a root, and this is obviously in $\mathbf{F}$.

Substituting the given root into the equation, we obtain:

$$(a + b\sqrt{k})^3 - 3(a + b\sqrt{k}) - 1 = 0$$

$$a^3 + 3a^2b\sqrt{k} + 3ab^2k + b^3k\sqrt{k} - 3a - 3b\sqrt{k} - 1 = 0$$

$$[a^3 + 3ab^2k - 3a - 1] + [3a^2b + b^3k - 3b]\sqrt{k} = 0.$$

But now we can conclude that $[3a^2b + b^3k - 3b] = 0$, for otherwise we could solve for $\sqrt{k}$ as an element of $\mathbf{F}$. But this then also requires that $[a^3 + 3ab^2k - 3a - 1] = 0$. After dividing the first of these two equations by b (which is not 0), we have

$$3a^2 + b^2k - 3 = 0$$

$$a^3 + 3ab^2k - 3a - 1 = 0.$$

Solving the first for b^2k and substituting the result into the second, we obtain

$$a^3 + 3a(3 - 3a^2) - 3a - 1 = 0$$

$$-8a^3 + 6a - 1 = 0$$

$$(-2a)^3 - 3(-2a) - 1 = 0$$

from which we see that $-2a$, which is in $\mathbf{F}$, is a solution of the original equation. ∎

Now the main result is easily accessible.

THEOREM 3. *It is not possible to trisect an arbitrary angle.*

Proof. If it were possible to trisect an arbitrary angle, it would be possible to trisect, say, a 60° angle, which is itself constructible. But, as noted earlier, this would imply the constructibility of cos 20° and hence the constructibility of a root of the cubic equation $x^3 - 3x - 1 = 0$. By Theorem 1, such a root would have to belong to a field $\mathbf{F}_N$ situated in a sequence $\mathbf{Q} = \mathbf{F}_0 \subset \mathbf{F}_1 \subset \mathbf{F}_2 \subset \cdots \subset \mathbf{F}_N$ in which each field is a quadratic extension of the previous one. By a repeated application of Lemma 3, then, there would have to be a *rational* root of $x^3 - 3x - 1 = 0$. But this will now be shown to be impossible.

If a rational number M/N, expressed in lowest terms, were a root of the equation, we would have

$$\frac{M^3}{N^3} - \frac{3M}{N} - 1 = 0,$$

and so

$$M^3 - 3MN^2 - N^3 = 0.$$

Thus we could write both

$$M^3 = N[3MN + N^2]$$

and

$$N^3 = M[M^2 - 3N^2].$$

From the first we conclude that if N had a prime factor p, then p would divide M^3 and hence M. From the second we conclude that if M had a prime factor q, then q would divide N^3 and hence N. Since M and N were assumed to have no factors in common, the only possible values for M and N are ± 1, and so the only possible rational roots are ± 1. But by inspection, neither of these actually turns out to be a root. ∎

It is important to realize that there is a difference between the questions:

(i) Is it possible to construct the angle θ?

(ii) Given the angle 3θ, is it possible to construct the angle θ?

The second question is the one addressed by the angle trisection problem, and it only reduces to the first question in cases where the angle 3θ is itself constructible. It happened to be convenient to use such a case in the proof of Theorem 3, but the reader should know that there are angles for which the answer to question (i) is "no" but the answer to question (ii) is "yes". To analyze these two different questions, it is useful to convert them into more algebraic terms. Remembering that in the presence of a unit segment we have available an angle if and only if we have available a segment whose length is its cosine, we can rephrase these questions as follows:

(i′) Is $\cos\theta$ constructible?

(ii′) Is $\cos\theta$ constructible if, in addition to a unit segment, we are also given a segment of length $\cos 3\theta$?

In Problem 4, the reader is asked to show the existence of an angle for which these questions have different answers.

PROBLEMS

1. It is obvious that by repeated bisection it is possible to divide an arbitrary angle into 4 equal parts. Show

how this may also be deduced algebraically from the equation relating $\cos 4\theta$ and $\cos\theta$.

2. Discuss the validity of the following proof of Theorem 3: "By the continuity of $\cos\theta$, there exists an angle θ_0 such that $\cos\theta_0 = (\sqrt[3]{2})/2$. By Theorem 2, $\cos\theta_0$ is not constructible. Hence $3\theta_0$ is an angle that cannot be trisected."

*3. Prove that it is not possible, in general, to quintisect an arbitrary angle, i.e., to divide it into five equal parts.

4. Is the difficulty in Problem 2 real or imagined? Show that there are trisectible angles 3θ such that θ is not constructible.

Section 1.4. Squaring the Circle

The third classical construction problem, that of squaring the circle, turns out to be much more difficult than the previous two. The question is: **Given a circle, is it possible to construct a square with the same area as the circle**? If we take as our unit length the radius of the circle, its area is π, and so the construction of the appropriate square amounts to the construction of a segment of length $\sqrt{\pi}$, which would necessitate the constructibility of π itself. It will be shown, however, that π is not constructible, and thus that it is not possible to square the circle. The proof that π is not constructible will consist of two main parts: first, that every constructible number is the root of some polynomial equation with rational coefficients; second, that π cannot be a root of any polynomial equation with rational coefficients. The first part, which is fairly simple, will be covered in this section. The second part, which is much more difficult, will be proved in Section 1.7, the intervening sections being used to introduce some ideas needed there as well as later on.

We shall make use of the following simple lemma:

LEMMA 4. *Let* $\mathbf{F}(\sqrt{k})$ *be a quadratic extension of a field* $\mathbf{F}$. *If a is a root of a polynomial equation of degree n with coefficients in* $\mathbf{F}(\sqrt{k})$, *then a is a root of a polynomial equation of degree* $2n$ *with coefficients in* $\mathbf{F}$.

Proof. By hypothesis, a is a root of an equation of the form

$$\left(a_n + b_n\sqrt{k}\right)x^n + \left(a_{n-1} + b_{n-1}\sqrt{k}\right)x^{n-1} + \cdots + \left(a_0 + b_0\sqrt{k}\right) = 0,$$

where for all i, a_i and $b_i \in \mathbf{F}$, and furthermore, a_n and b_n are not both zero. Transferring the terms involving $\sqrt{k}$ to the other side of the equation, we obtain

$$a_n x^n + a_{n-1}x^{n-1} + \cdots + a_0 = -\sqrt{k}\left(b_n x^n + b_{n-1}x^{n-1} + \cdots + b_0\right).$$

Upon squaring both sides and collecting all terms together again on the left, we have a polynomial equation of which a is also a root. Its degree is $2n$, because the coefficient of x^{2n}, the highest power present, is $a_n^2 - kb_n^2$, which is not 0 by the assumption that $\sqrt{k} \notin \mathbf{F}$. ∎

Using this lemma, we are now able to prove:

THEOREM 4. *Every constructible number is the root of some polynomial equation with rational coefficients.*

Proof. If the number a is constructible, then by Theorem 1 there is a sequence of fields $\mathbf{Q} = \mathbf{F}_0 \subset \mathbf{F}_1 \subset \cdots \subsetneq \mathbf{F}_N$ such that $a \in \mathbf{F}_N$ and for each j, $0 \leqslant j$

$\leqslant N-1$, $\mathbf{F}_{j+1}$ is a quadratic extension of $\mathbf{F}_j$. Now, a satisfies a polynomial equation of degree 1 with coefficients in $\mathbf{F}_N$, namely, the equation $x-a=0$. By the repeated application of Lemma 4, a satisfies a polynomial equation of degree 2^i with coefficients in $\mathbf{F}_{N-i}$. Hence for $i=N$, we see that a satisfies a polynomial equation of degree 2^N with coefficients in $\mathbf{Q}$. ∎

In view of Theorem 4, once it has been shown (in Section 1.7) that π is not the root of any polynomial equation with rational coefficients, we shall have a complete proof of the result stated earlier, namely:

THEOREM 5. *It is not possible to "square the circle".*

PROBLEMS

1. The number $\sqrt{2}+\sqrt{3}$, being constructible, has associated with it a sequence of quadratic field extensions $\mathbf{Q}=\mathbf{F}_0 \subset \mathbf{F}_1 \subset \cdots \subset \mathbf{F}_N$ such that $\sqrt{2}+\sqrt{3} \in \mathbf{F}_N$. Determine the smallest such number N. Using this value, what is the degree of the polynomial whose existence is proved in Theorem 4? Following the procedure in the proof of Theorem 4, determine this and any intermediate polynomials.

2. Does there exist a polynomial equation with rational coefficients and of degree less than that determined in Problem 1 such that $\sqrt{2}+\sqrt{3}$ is a root?

Section 1.5. Polynomials and Their Roots

It now becomes essential for us to expand our considerations to the set $\mathbf{C}$ of complex numbers. These are, of course, numbers of the form $a + bi$, where a and b belong to **R** and $i = \sqrt{-1}$. In fact, we may regard **C** as a quadratic extension of **R**, $\mathbf{C} = \mathbf{R}(\sqrt{-1})$, now that $\sqrt{-1}$ is in the set we are considering. The complex numbers form a field, which can be shown by verifying that the proof of Lemma 1c is also valid for $k < 0$; in this case $k = -1$.

From now on, the term *field* will refer to *a subset of* **C** *containing the number* 1 *and closed under the rational operations*, which definition certainly includes all the fields we have been considering so far. By a *polynomial over a field* **F** is meant a function of the form $a_n x^n + a_{n-1}x^{n-1} + \cdots + a_1 x + a_0$ where the coefficients a_i are elements of **F** and where n is a non-negative integer. The domain of the independent variable x will be understood to be **C** except where stated otherwise. The *degree* of such a polynomial is the highest exponent of x in a nonzero term. The degree of the polynomial identically equal to 0 is undefined. We shall represent polynomials by conventional function notation. By a *root* of a polynomial is meant a value of x for which the function value is 0.

The following theorem, which we will not prove, is called the **Fundamental Theorem of Algebra**:

THEOREM 6. *Every polynomial over* **C** *of degree* $\geqslant 1$ *has at least one root* (*in* **C**).

The proof of this theorem is most easily accomplished using elementary complex variable theory, and the reader may find it in almost any book on that subject. More elementary proofs are cited in the "References and Notes" section at the end of this chapter. Included as special cases, of course, are polynomials over **R** or over **Q**.

If $f(x)$ and $g(x)$, which we may also refer to simply as f and g, are polynomials, we say that $g(x)$ *divides* $f(x)$ if there is a polynomial $q(x)$ such that $f(x) = g(x)q(x)$. If $f(x)$ and $g(x)$ are polynomials over a field **F**, then by the usual long division of polynomials as in high school algebra, it follows that $q(x)$ is also over **F**. Clearly, for g to divide f, the degree of g must be less than or equal to the degree of f.

It is an elementary observation that if f is a polynomial, the number r is a root of f if and only if $(x - r)$ divides f. For, as long division shows, the remainder upon dividing f by $(x - r)$ is simply $f(r)$. (See Problem 1.) If r is a root of f, we define the *multiplicity of the root* r to be the largest integer m such that $(x - r)^m$ divides f. This leads to the following refinement of Theorem 6:

THEOREM 7. *Every polynomial over* **C** *of degree n has exactly n roots, each counted as many times as its multiplicity.*

Proof. By induction on n. When $n = 0$, the polynomial is a nonzero constant, and hence has no roots. If f is a polynomial of degree $n \geqslant 1$, then by Theorem 6 it has a root r. Factoring $(x - r)$ out of $f(x)$, we are left with a polynomial of degree $n - 1$, to which the inductive hypothesis applies, giving another $n - 1$ roots (some of which might coincide with r or with each other). ∎

Most of the material encountered so far in this section is probably somewhat familiar to the reader. But now we want to take a closer look at certain sets of numbers appearing as the roots of polynomials. In particular, we say that a number a (which, as usual, may be complex) is *algebraic over the field* **F** if there is some polynomial (not identically 0) over **F** of which a is a root. *In cases where* **F** *is not specified, we understand it to be* **Q**. Thus we would say that 3, i, and $\sqrt[3]{7}$ are algebraic, as they are roots, respectively, of the following polynomials over $\mathbf{Q}$: $x-3$, x^2+1, x^3-7. Any number which is not algebraic is called *transcendental*.

To phrase our current problem in this language, we have shown (Section 1.4) that *all constructible numbers are algebraic*. To complete the proof of the impossibility of squaring the circle, we want to show that *π is transcendental*.

Let us now make a very important observation, which is not at all obvious:

THEOREM 8. *If* $\mathbf{F}$ *is a field, then the set of all numbers algebraic over* $\mathbf{F}$ *also forms a field.*

Proof. Suppose that a and b are algebraic over $\mathbf{F}$, and thus they satisfy, respectively, polynomial equations

$$a_n x^n + a_{n-1} x^{n-1} + \cdots + a_0 = 0,$$

$$b_m x^m + b_{m-1} x^{m-1} + \cdots + b_0 = 0,$$

where all the coefficients are in $\mathbf{F}$ and $a_n, b_m \neq 0$. It is easy to see that $-b$ is algebraic over $\mathbf{F}$, as it satisfies

$$(-1)^m b_m x^m + (-1)^{m-1} b_{m-1} x^{m-1} + \cdots + b_0 = 0;$$

and if $b \neq 0$, $1/b$ is algebraic over $\mathbf{F}$, being a root of

$$b_m + b_{m-1} x + \cdots + b_0 x^m = 0.$$

Thus, it suffices to show that $a+b$ and ab are algebraic over $\mathbf{F}$, because subtraction and division reduce to these cases. One more observation: using the original equations, a^n and b^m can be expressed in terms of lower powers of a and b respectively; for example, $a^n = -(a_{n-1} a^{n-1} + \cdots + a_0)/a_n$.

Now consider the table of mn numbers:

$$\begin{array}{ccccc} 1 & a & a^2 & \cdots & a^{n-1} \\ b & ab & a^2b & \cdots & a^{n-1}b \\ b^2 & ab^2 & a^2b^2 & \cdots & a^{n-1}b^2 \\ \vdots & & & & \\ b^{m-1} & ab^{m-1} & a^2b^{m-1} & \cdots & a^{n-1}b^{m-1} \end{array}.$$

Listing them in any order, and calling them c_1, c_2, $\ldots c_R$, where $R = mn$, we note that for every i, ac_i can be written as a linear combination of all the c_i's:

$$ac_i = d_{i1}c_1 + d_{i2}c_2 + \cdots + d_{iR}c_R,$$

where the coefficients $d_{ij} \in \mathbf{F}$. Actually, most of the d_{ij}'s are 0; for example if c_i is one of the entries not in the last column of the table, ac_i is just the single entry to its right. If c_i is one of the entries in the last column, then ac_i has the form $a^n b^k$, in which a^n is then replaced by its equivalent in terms of lower powers of a. In any case, the above representation is possible. Denoting by C the column vector whose R entries are the c_i's, and by D the $R \times R$ matrix of the d_{ij}'s, we can rewrite the above equations in matrix form,

$$aC = DC.$$

Similarly, there is a matrix E such that

$$bC = EC.$$

Thus we have

$$(a + b)C = (D + E)C.$$

Since $C \neq 0$, this implies that $a + b$ is an eigenvalue of the matrix $D + E$. But then $a + b$ is a root of the characteristic polynomial, $\det(D + E - xI)$, where I is the identity matrix. Since this is a polynomial over the field **F**, we conclude that $a + b$ is algebraic over **F**.

For the product ab, observe that

$$(ab)C = a(bC) = a(EC) = E(aC) = E(DC) = (ED)C,$$

and thus ab is an eigenvalue of ED. So by the same reasoning as above, ab is algebraic over **F**. ∎

It is useful to note that if f is a polynomial over a field **F**, division of f by its *leading coefficient* (the coefficient of the highest power of x) yields another polynomial over **F** of the same degree and with the same roots as f. The new polynomial is *monic*, meaning that its leading coefficient is 1. It is easy to see that a monic polynomial can be written as the product $\prod_i (x - r_i)$, where the r_i's are its roots, each occurring as often as its multiplicity. By a similar procedure, for any polynomial over **Q** there can be found a polynomial with integral coefficients having the same degree and roots as f.

PROBLEMS

1. a. If f is a polynomial of degree $n \geqslant 1$ over a field **F**, and if r is any element of **F**, show that there is a polynomial q of degree $n - 1$ over **F** and a number R in **F** such that $f(x) = (x - r)q(x) + R$.

   b. Show that, in the above, $R = f(r)$.

   c. Let f be a polynomial over **C**. Show that r is a root of f if and only if $(x - r)$ is a factor of f.

2. If f is a polynomial with integral coefficients, it is sometimes useful to know whether it has any rational roots. Suppose M/N is a rational root expressed in lowest

terms. There is a simple necessary condition connecting M and N with the first and last coefficients of the polynomial. Find it. (This result is called the **Rational Roots Theorem.** It gives a necessary condition that can be used to narrow down the candidates for a rational root to a small collection which can each be tested.)

3. Use the result of Problem 2 to find any rational roots of each of the following:
   a. $8x^3 - 6x - 1$
   b. $x^3 - 3x - 1$
   c. $2x^3 + 3x^2 - x - 1$
   d. $3x^5 - 5x^3 + 5x - 1$
   e. $x^3 + 4x^2 + x - 6$.
4. If $\mathbf{F}$ is a countable field, show that the field of all numbers algebraic over $\mathbf{F}$ is also countable.
5. Show that the result of Problem 2 of Section 1.1 is just a special case of Problem 2 of this section.
6. If r is a root of multiplicity $\geqslant 2$ of a polynomial f, show that $f'(r) = 0$, where f' denotes the derivative of f.

Section 1.6. Symmetric Functions

A very useful tool in the study of polynomial equations is the concept of a symmetric polynomial in several variables. To begin, a *polynomial over a field* $\mathbf{F}$ *in the variables* $x_1, x_2, \ldots, x_n$ is a function of the form

$$P(x_1, x_2, \ldots, x_n) = \sum_{i_1, i_2, \ldots, i_n} a_{i_1, i_2, \ldots, i_n} x_1^{i_1} x_2^{i_2} \cdots x_n^{i_n}$$

where the i's are non-negative integers, the coefficients, the a's, are elements of $\mathbf{F}$, and the sum contains only a finite number of terms. *The degree of a term* in such a polynomial is the sum of the exponents of the x_i's in the term, and the *degree of the polynomial* is the highest degree of any term. As in the case of polynomials in one variable, degree is not defined for the polynomial which is identically 0.

As examples of such polynomials, $x_1^2 + x_1x_2^3$ is a polynomial over $\mathbf{Q}$ in two variables and of degree 4; $\sqrt{2}\ x_1x_2x_3$ is a polynomial over $\mathbf{Q}(\sqrt{2})$ in three variables and of degree 3; $\sum_{i=1}^{n} x_i$ is a polynomial over $\mathbf{Q}$ in n variables and of degree 1.

Some polynomials in several variables have a curious property; they remain unchanged by any permutation of the variables! For example, $x_1^2 + x_2^2$ has this property, because if x_1 and x_2 are interchanged, the result is $x_2^2 + x_1^2$, which is the same polynomial that we started with. Similarly, $x_1x_2x_3$ also has this property; it is unchanged by any permutation of the variables. However $x_1^2 - x_2^2$ does not have this property; the interchange of x_1 and x_2 results in the negative of the original polynomial. We say that a polynomial in n variables is *symmetric* if it is unchanged by each of the $n!$ permutations of the variables. That is, if ϕ is a one-to-one mapping from the set of integers $1, 2, \ldots, n$ to itself (a *permutation of the first n integers*), for P to be symmetric we require that $P(x_{\phi(1)}, x_{\phi(2)}, \ldots, x_{\phi(n)}) = P(x_1, x_2, \ldots, x_n)$. So we would say that $x_1^2 + x_2^2$ is a symmetric polynomial in two variables, whereas $x_1^2 - x_2^2$ is not.

Certain important symmetric polynomials in n variables arise as the coefficients of the powers of x in the expansion of $\prod_{i=1}^{n}(x - x_i)$. Writing this expansion as a polynomial in x,

$$\prod_{i=1}^{n}(x - x_i) = x^n - \sigma_1 x^{n-1} + \sigma_2 x^{n-2} - \cdots + (-1)^n \sigma_n,$$

it is not hard to see that

$$\sigma_1 = x_1 + x_2 + \cdots + x_n$$
$$\sigma_2 = x_1x_2 + x_1x_3 + \cdots + x_2x_3 + \cdots + x_{n-1}x_n$$
$$\vdots$$
$$\sigma_i = \text{sum of all products of } i \text{ different } x_j\text{'s}$$
$$\vdots$$
$$\sigma_n = x_1x_2 \cdots x_n.$$

These σ_i's are obviously symmetric polynomials in $x_1, x_2, \ldots, x_n$. From these calculations we are able to make some useful observations about the roots of an nth degree polynomial in x. If $r_1, r_2, \ldots, r_n$ are the roots of $a_nx^n + a_{n-1}x^{n-1} + \cdots + a_0$, then the sum of the roots equals $-a_{n-1}/a_n$ and their product equals $(-1)^n a_0/a_n$, since

$$\prod_{i=1}^{n}(x - r_i) = x^n + \frac{a_{n-1}}{a_n}x^{n-1} + \frac{a_{n-2}}{a_n}x^{n-2} + \cdots + \frac{a_0}{a_n}.$$

(Elementary applications of these facts are pursued in the problems.) Symbolically, of course, the sum of the roots is $\sigma_1(r_1, r_2, \ldots, r_n)$ and their product is $\sigma_n(r_1, r_2, \ldots, r_n)$.

The σ_i's just defined are called the *elementary symmetric functions*, and it may be surprising to learn that every symmetric polynomial in n variables can be expressed as a polynomial in these elementary symmetric functions! This result, which will be proved below, is called the **Fundamental Theorem on Symmetric Functions**. As an example, we observe in the case $n = 3$ that

$$x_1^2 + x_2^2 + x_3^2 = (x_1 + x_2 + x_3)^2 - 2(x_1x_2 + x_1x_3 + x_2x_3)$$
$$= \sigma_1^2 - 2\sigma_2.$$

THEOREM 9. *Every symmetric polynomial P over $\mathbf{F}$ in $x_1, x_2, \ldots, x_n$ can be written as a polynomial Q over $\mathbf{F}$ in the elementary symmetric functions. If P has all integral coefficients, then so too does Q. The degree of Q is less than or equal to the degree of P.*

Proof. Let M be arbitrary but fixed, $M \geqslant 1$. We shall prove the result for all polynomials in $x_1, x_2, \ldots, x_n$ of degree less than or equal to M.

To begin, we associate with each term of the symmetric polynomial P the n-tuple containing the exponents of $x_1, x_2, \ldots, x_n$ respectively (some of which may be 0). A linear ordering of these n-tuples can be obtained by defining $(i_1, i_2, \ldots, i_n) > (j_1, j_2, \ldots, j_n)$ if in the first position from the left in which they differ, say position k, $i_k > j_k$. (This is the same idea as alphabetical ordering.) With respect to this ordering, the term of P with the highest n-tuple of exponents is called the *highest term.* (This can be different from the term of highest degree.) For all such polynomials of degree $\leqslant M$, there are clearly only a finite number of possible exponent n-tuples less than the exponent n-tuple of the highest term. We denote this number by N, and our proof will be by induction on N.

For example, if $M = 4$ and $P(x_1, x_2, x_3) = x_1^3 + x_2^3 + x_3^3$, then $N = 31$, for the highest exponent 3-tuple is (3,0,0), and this may be followed, in decreasing order among all possible 3-tuples in polynomials of degree $\leqslant 4$, by: (2, 2, 0), (2, 1, 1), (2, 1, 0), (2, 0, 2), (2, 0, 1), (2, 0, 0),

(1, 3, 0), (1, 2, 1), (1, 2, 0), . . . , (0, 0, 1), (0, 0, 0), of which there are a total of 31.

To prove the result for $N = 0$, we observe that in this case the polynomial is a nonzero constant, which can also be considered as a (constant) polynomial in the σ_i's.

Assuming now that the result is true for $N \leqslant R$, we shall prove it for $N = R + 1$. If the highest term of P is $ax_1^{i_1}x_2^{i_2} \cdots x_n^{i_n}$, we first observe that $i_1 \geqslant i_2 \geqslant \cdots \geqslant i_n$. For, if there were a j with $i_j < i_{j+1}$, the permutation which simply interchanges x_j and x_{j+1} would result in a higher term, which, since P is symmetric, must be included in P. Thus we can construct the polynomial $Q = a\sigma_1^{i_1-i_2}\sigma_2^{i_2-i_3} \cdots \sigma_{n-1}^{i_{n-1}-i_n}\sigma_n^{i_n}$, which is a polynomial in the σ_i's of degree less than or equal to the degree of P. If for each σ_i we substitute its expression in $x_1, x_2, \ldots, x_n$, and then expand out in the form of a polynomial in the x_i's, we find that the highest term is precisely the same as the highest term of P. To see this, observe that the highest term of a product of polynomials is the product of their highest terms (see Problem 5). Thus the highest term in the expansion of Q is $ax_1^{i_1-i_2}(x_1x_2)^{i_2-i_3} \cdots (x_1x_2 \cdots x_n)^{i_n} = ax_1^{i_1}x_2^{i_2} \cdots x_n^{i_n}$. It is also easy to see that every term of this expansion has the same degree, and thus, as a function of the x_i's, Q has degree less than or equal to the degree of P. Consequently, the difference $\tilde{P} \equiv P - Q$ is either 0, in which case we are done, or else another symmetric polynomial with a lower highest term. In this case, the inductive hypothesis applies to $\tilde{P}$, and so $P = \tilde{P} + Q$ which is of the required form. If P has integral coefficients, then so too does $\tilde{P}$. ∎

An important consequence of Theorem 9 which will be applied repeatedly in our treatment of the roots of polynomials is the following:

COROLLARY 9a. *Let $f(x)$ be a polynomial over* **F** *of degree n with roots $r_1, r_2, \ldots, r_n$. If P is any symmetric polynomial over* **F** *in n variables, then $P(r_1, r_2, \ldots, r_n)$ is an element of* **F**.

Proof. By Theorem 9, P may be expressed as a polynomial in the elementary symmetric functions. But if $f(x) = a_n x^n + a_{n-1} x^{n-1} + \cdots + a_0$, then $\sigma_i(r_1, r_2, \ldots, r_n) = \pm a_{n-i}/a_n$, which is in **F**. Hence $P(r_1, r_2, \ldots, r_n)$ is itself in **F**, since **F** is closed under the operations used to evaluate polynomials (addition, subtraction, and multiplication). ∎

As an example of this result, consider the polynomial $f(x) = x^2 + 2x - 1$, which is over **Q**, and which has as roots $r_1 = -1 + \sqrt{2}$ and $r_2 = -1 - \sqrt{2}$. Taking any symmetric polynomial in two variables, say $P(x_1, x_2) = x_1^3 + x_2^3$, the corollary asserts that $P(r_1, r_2)$ will turn out to be rational. We can verify this directly:

$$
\begin{aligned}
P(r_1, r_2) &= (-1 + \sqrt{2})^3 + (-1 - \sqrt{2})^3 \\
&= (-1 + 3\sqrt{2} - 6 + 2\sqrt{2}) \\
&\quad + (-1 - 3\sqrt{2} - 6 - 2\sqrt{2}) \\
&= -14,
\end{aligned}
$$

which is indeed in **Q**.

Using Corollary 9a, it is possible to give another proof that the set of all numbers which are algebraic over one field is a field itself (Theorem 8). This is pursued in Problem 6. Corollary 9a also leads to the more general result, Corollary 9b, which is somewhat technical but forms an essential link in our upcoming proof that π is transcendental.

COROLLARY 9b. *Let f be a polynomial of degree n over* **F** *with roots* $r_1, r_2, \ldots, r_n$. *Let R be fixed,* $1 \leqslant R \leqslant n$. *Let* $s_1, s_2, \ldots, s_m$ *be the collection of all sums* $r_{i_1} + r_{i_2} + \cdots + r_{i_R}$, *where the R subscripts are all different. Then there is a polynomial over* **F** *whose roots are precisely* $s_1, s_2, \ldots, s_m$.

Proof. The candidate is certainly $\prod_{i=1}^{m}(x - s_i)$; one merely has to show that if this is expanded out, the coefficient of each power of x is an element of **F**. Since the coefficients are, except possibly in sign, the elementary symmetric functions of m variables, evaluated at the s_i's, it will certainly suffice to show that *any* symmetric polynomial over **F** in m variables, evaluated at the s_i's, is an element of **F**. Let $P(x_1, x_2, \ldots, x_m)$ be such a symmetric polynomial. We consider n new variables, $u_1, u_2, \ldots, u_n$, and we replace in P each x_i by a certain specific sum of R different u_i's, each x_i being replaced by a different such sum. Then we claim that P becomes a symmetric polynomial over **F** in the variables $u_1, u_2, \ldots, u_n$. To see this, we first note that any permutation of the u_i's may be accomplished by a sequence of *transpositions* (single interchanges of two of the u_i's). See Problem 7 for a proof. Thus it remains to show that P is invariant under the transposition interchanging, say, u_i and u_j. In this case the x_i's containing neither or both of these as summands are unchanged, and the remainder are permuted. Thus, by the symmetry of P in the x_i's, P is left unchanged. Plugging in the r_i's for the u_i's, we get $P(s_1, s_2, \ldots, s_m)$, which by Corollary 9a is an element of **F**. ∎

PROBLEMS

1. The cubic $2x^3 - 3x^2 - 32x - 15$ has two roots whose sum is 2. Find all the roots.

2. Find the product of all the roots of $2x^5 - 3x^4 + 7x^3 - 7x^2 + 3x - 1$. Are any of the roots rational?

3. Find the value, expressed in terms of the coefficients, of the sum of the squares of the roots of $a_n x^n + a_{n-1}x^{n-1} + \cdots + a_0$.

4. Find the value, expressed in terms of the coefficients, of the sum of the reciprocals of the roots of $a_n x^n + a_{n-1}x^{n-1} + \cdots + a_0$, where $a_0 \neq 0$.

5. Let P_1 and P_2 be two polynomials in n variables. Show that the highest term of the product $P_1 P_2$ is the product of the highest terms of P_1 and P_2. (The word "highest" refers to the ordering on the exponent n-tuples.)

6. (Alternate proof of Theorem 8.) Suppose a and b are algebraic over a field **F**. Then a is a root of some polynomial over **F**, whose complete set of roots may be written $a = a_1, a_2, \ldots, a_n$, where each root is listed as many times as its multiplicity. Similarly for b, giving rise to roots $b = b_1, b_2, \ldots, b_m$. Apply the theory of symmetric functions twice successively to the expression

$$\prod_{i=1}^{n} \prod_{j=1}^{m} (x - a_i - b_j)$$

to conclude that $a + b$ is algebraic over **F**. In a similar way, show that ab is also algebraic over **F**.

7. Prove that every permutation of the variables $x_1, x_2, \ldots, x_n$ can be accomplished by a succession of transpositions, where a transposition is the single interchange of two of the variables. (This was used in the proof of Corollary 9b.)

8. With reference to the preceding problem, there may be many different sequences of transpositions resulting in the same permutation. Show that for a given permutation, all sequences of transpositions leading to it have the same parity; that is, either they all contain an even number of transpositions or else they all contain an odd number. (In this way, permutations can be classified as "even" or "odd".)

Section 1.7. The Transcendence of π

The proof of the transcendence of π relies on a wide collection of mathematical techniques, ranging from calculus to elementary number theory. In this respect, in addition to the fundamental importance of the result, it should be of interest and worthy of careful study.

Use will be made of the function e^z for z complex, with which the reader may be unfamiliar; so we begin with some comments about this function. If z is a complex number it may be written in the form $z = x + iy$, where x and y are real. The numbers x and y are called the *real* and *imaginary* parts of z, respectively. The expression e^z is then defined to be the complex number $e^x(\cos y + i \sin y)$, which involves only the usual exponential and trigonometric functions of real variables. From the definition it follows that $e^{\pi i} = e^0(\cos \pi + i \sin \pi) = -1$. Also, it can be shown rather easily (Problem 1) that e^z obeys the usual laws of exponents; in particular, $e^{z_1}e^{z_2} = e^{z_1+z_2}$. These are the only properties that we shall use here.

One number theoretic lemma will also be useful:

LEMMA 10. *The product of any n consecutive positive integers is divisible by $n!$.*

Proof. Calling the integers $m+1, m+2, \ldots, m+n$, we wish to show that the quotient $(m+1)(m+2)\cdots(m+n)/n!$ is an integer. Multiplying it by 1, written as $m!/m!$, puts it in the form $(m+n)!/m!\,n!$, which is simply one of the binomial coefficients, and hence an integer. (Cf. Problem 2.) ∎

We are now ready to prove:

THEOREM 10. *π is transcendental.*

Proof. Suppose π is algebraic. This will eventually lead to a contradiction. Since i is also algebraic, πi is algebraic by Theorem 8, and thus there is a polynomial over $\mathbf{Q}$ whose roots are $\pi i = r_1, r_2, \ldots, r_n$. Since $e^{r_1} = -1$, we have

$$(e^{r_1}+1)(e^{r_2}+1)\cdots(e^{r_n}+1) = 0,$$

which we can multiply out, combining only the terms with 0 exponents, to obtain

$$e^{q_1} + e^{q_2} + \cdots + e^{q_m} + k = 0$$

where k is a positive integer, being a sum of 1's, and the q_i's, some of which may be equal, are all different from 0.

Claim 1. There is a polynomial with integral coefficients whose roots are precisely $q_1, q_2, \ldots, q_m$.

Proof. In expanding out the product as described above, the set of terms we obtain before combining anything consists of the single number 1 and the number e raised to various powers. The entire collection of such powers is just the set of all sums of R distinct r_i's (distinct by subscript at least), as R ranges from 1 to n. By Corollary 9b, for each fixed value of R, there is a polynomial over $\mathbf{Q}$ with precisely those sums as roots. The product over all R of these polynomials is practically what we want. We simply need to modify it slightly first by dividing through by the appropriate power of x so as to eliminate the 0 roots, and then by multiplying through by an appropriate integer to make all its coefficients integral.

We shall call this polynomial g, and we shall write it as $g(x) = ax^m + a_{m-1}x^{m-1} + \cdots + a_0$. (The subscript has been left off the leading coefficient for the sake of convenience.) Its roots are precisely $q_1, q_2, \ldots, q_m$.

Now define an auxiliary polynomial f by

$$f(x) = \frac{a^s x^{p-1}[g(x)]^p}{(p-1)!},$$

where p is a prime to be chosen appropriately later and $s = mp - 1$. The degree of f is $p - 1 + pm = s + p$. The polynomial f is a complex-valued function of a complex variable, and it is necessary for us to make use of some of the differentiation theory for such functions. The results needed are formally identical with results from elementary calculus, but of course they require separate justification. These results will be summarized in the text, their actual development being relegated to the problems.

For complex functions of a complex variable, derivatives are defined by the same kind of limit of a difference quotient as in the real case. The usual rules for the derivative of sums, differences, products, and quotients carry over, as do the formulas for the derivatives of polynomials and the exponential function. (See Problem 3.) In this context, denoting the jth derivative of f by $f^{(j)}$, we define

$$F(x) = f(x) + f^{(1)}(x) + f^{(2)}(x) + \cdots + f^{(s+p)}(x).$$

Since $f^{(s+p+1)}(x) = 0$, we find that the derivative of $e^{-x}F(x)$ is just $-e^{-x}f(x)$, for

$$\begin{aligned}\frac{d}{dx}\left[e^{-x}F(x)\right] &\\ &= e^{-x}\left[f^{(1)}(x) + f^{(2)}(x) + \cdots + f^{(s+p)}(x)\right]\\ &\qquad - e^{-x}\left[f(x) + f^{(1)}(x) + f^{(2)}(x) + \cdots + f^{(s+p)}(x)\right]\\ &= -e^{-x}f(x).\end{aligned}$$

This fact will be used in what follows.

Next let us regard x as an arbitrary but fixed complex number, and look at the function given by

$$\phi(u) = e^{-xu}F(xu),$$

where u is a *real* variable. This is a complex-valued function of a real variable. Differentiation and integration of such functions are defined by applying these operations separately to the real and imaginary parts, both of which are simply real functions of a real variable. It is very simple to verify that the Fundamental Theorem of Calculus, in the form

$$\phi(b) - \phi(a) = \int_a^b \phi'(u)\,du,$$

carries over to such functions (Problem 4).

It is our intention to apply this theorem on the interval [0, 1]. To this end, we need to calculate $\phi'(u)$. This may be done by a version of the chain rule whose justification in the present context is treated in Problem 5. Using the intermediate variable

$$z = ux$$

we may write

$$\phi(u) = e^{-z}F(z).$$

It follows that

$$\begin{aligned}\phi'(u) &= \frac{d}{dz}\left[e^{-z}F(z)\right]\cdot\frac{dz}{du}\\ &= -e^{-z}f(z)\cdot x\\ &= -xe^{-ux}f(ux).\end{aligned}$$

Now, from the Fundamental Theorem of Calculus, we obtain:

$$\phi(1) - \phi(0) = \int_0^1 \phi'(u)du,$$

$$e^{-x}F(x) - e^{-0}F(0) = \int_0^1 -xe^{-ux}f(ux)du,$$

$$F(x) - e^{x}F(0) = \int_0^1 -xe^{(1-u)x}f(ux)du.$$

Letting x successively assume the values $q_1, q_2, \ldots, q_m$ and then adding the results, we have

$$\sum_{j=1}^{m} F(q_j) + kF(0) = -\sum_{j=1}^{m}\int_0^1 q_je^{(1-u)q_j}f(uq_j)du.$$

Eventually we shall establish the desired contradiction by choosing the prime p so as to make the left side of this equation a nonzero integer and the right side arbitrarily small.

Claim 2. $\sum_{j=1}^{m} F(q_j)$ is an integer divisible by p.

Proof. Using the definition of F, we have

$$\sum_{j=1}^{m} F(q_j) = \sum_{j=1}^{m} \sum_{r=0}^{s+p} f^{(r)}(q_j) = \sum_{r=0}^{s+p} \sum_{j=1}^{m} f^{(r)}(q_j).$$

For $0 \leqslant r < p$,

$$\sum_{j=1}^{m} f^{(r)}(q_j) = 0,$$

since by the definition of f, $f^{(r)}(q_j)$ has at least one factor $g(q_j)$, which is 0. Thus,

$$\sum_{j=1}^{m} F(q_j) = \sum_{r=p}^{s+p} \sum_{j=1}^{m} f^{(r)}(q_j).$$

Now let r be arbitrary but fixed, $p \leqslant r \leqslant s+p$. Since $a^{-s}(p-1)!f(x)$ is a polynomial with integral coefficients, the coefficient of each nonzero term of its rth derivative contains the product of r (and hence at least p) consecutive integers; therefore, by Lemma 10, it is divisible by $p!$. Thus each coefficient of $f^{(r)}(x)$ is divisible by pa^s, which implies that the expression

$$\sum_{j=1}^{m} \frac{1}{pa^s} f^{(r)}(u_j)$$

is a symmetric polynomial with integral coefficients in the m variables $u_1, u_2, \ldots, u_m$. By the Fundamental Theorem on Symmetric Functions it may be expressed as a polynomial in the elementary symmetric functions and having integral coefficients. We now substitute in for the variables the values $q_1, q_2, \ldots, q_m$, and we recall that each elementary symmetric function $\sigma_i(q_1, q_2, \ldots, q_m) = \pm a_{m-i}/a$. Since the degree of the polynomial in the σ_i's is $\leqslant s$, each term has the form of an integer divided by some power of a not greater than s. Thus

$$\sum_{j=1}^{m} f^{(r)}(q_j) = p\left[a^s \sum_{j=1}^{m} \frac{1}{pa^s} f^{(r)}(q_j)\right],$$

where the bracketed term is an integer. Since this holds for each r such that $p \leqslant r \leqslant s+p$, the sum over those r is an integer divisible by p. Since this sum equals the original expression, the proof of the claim is complete.

Claim 3. For p sufficiently large, $kF(0)$ is an integer not divisible by p.

Proof. The terms of $F(0)$ fall into three categories. The terms $f^{(r)}(x)$ for $0 \leqslant r \leqslant p-2$ all contain a factor x, and hence are all 0 at $x = 0$. The terms $f^{(r)}(x)$ for $p \leqslant r \leqslant s+p$ are polynomials with integral coefficients each divisible by p, as shown in the proof of the previous claim. Thus $f^{(r)}(0)$, their constant terms, are integers divisible by p. For $kF(0)$ not to be divisible by p, therefore, it suffices to guarantee that the one remaining term, $kf^{(p-1)}(0)$, not be divisible by p. From the definition of f, the only term contributing to $f^{(p-1)}(0)$ is $(a^s a_0^p/(p-1)!)x^{p-1}$. Hence $kf^{(p-1)}(0) = ka^s a_0^p$. Consequently, if p is any prime greater than k, a, and a_0, it cannot divide this product.

Now we return to our earlier equation, in connection with which the two previous claims were undertaken:

$$\sum_{j=1}^{m} F(q_j) + kF(0) = -\sum_{j=1}^{m} \int_0^1 q_j e^{(1-u)q_j} f(uq_j)\,du.$$

We have shown (Claim 2) that the first term on the left is an integer divisible by p and (Claim 3) that for all p sufficiently large the second term on the left is an integer not divisible by p. Thus the left side is a nonzero integer. We will now show that for p sufficiently large, the right side may be made arbitrarily small, and hence not equal to any nonzero integer. This contradiction will complete the proof.

Now,

$$\left| -\sum_{j=1}^{m} \int_0^1 q_j e^{(1-u)q_j} f(uq_j)\,du \right|$$

$$= \left| \int_0^1 \sum_{j=1}^{m} q_j e^{(1-u)q_j} \left[\frac{a^s (uq_j)^{p-1} [g(uq_j)]^p}{(p-1)!} \right] du \right|$$

$$\leqslant \int_0^1 \sum_{j=1}^{m} \left| \frac{[a^m (uq_j) g(uq_j)]^{p-1}}{(p-1)!} q_j e^{(1-u)q_j} a^{m-1} g(uq_j) \right| du$$

since $a^s = a^{mp-1} = (a^m)^{p-1} a^{m-1}$. (The vertical bars on the right side of the inequality refer to the magnitudes, or *moduli*, of complex numbers. See Problem 3. The usual rules for absolute values of integrals and sums carry over directly to this case.) Let B be a uniform bound over $1 \leqslant j \leqslant m$ of the continuous functions $|a^m(uq_j)g(uq_j)|$ on the closed interval $0 \leqslant u \leqslant 1$. Let C be a bound on the continuous function

$$\sum_{j=1}^{m} |q_j e^{(1-u)q_j} a^{m-1} g(uq_j)|$$

on the same interval. Then the original expression is bounded by $CB^{p-1}/(p-1)!$. Since this is the pth term in the (convergent) Taylor series around 0 of Ce^B, it must approach 0 as $p \to \infty$. Thus it can be made arbitrarily small for p sufficiently large. ∎

As noted in Section 1.4, the proof of Theorem 10 also completes the proof that it is impossible to "square the circle".

PROBLEMS

1. Prove that $e^{z_1}e^{z_2} = e^{z_1+z_2}$.

2. Prove that the binomial coefficients are in fact integers.

3. The *modulus* of a complex number $z = x + iy$, where x and y are real, is given by $|z| = \sqrt{x^2 + y^2}$. If $F(z)$ is a complex function of a complex variable, then the definition of the statement $\lim_{z \to z_0} F(z) = L$ is that for every $\epsilon > 0$ there exists a $\delta > 0$ such that $|F(z) - L| < \epsilon$ whenever $0 < |z - z_0| < \delta$. This is formally identical with the real case, the difference being the use of moduli instead of absolute values. If $G(z)$ is a complex function of a complex variable, the derivative of G is the new function whose value at any point z_0 is given by

$$G'(z_0) = \lim_{z \to z_0} \frac{G(z) - G(z_0)}{z - z_0} .$$

(As in the real case, some complex functions are differentiable and some are not.) Show that the usual rules from elementary calculus carry over in the complex case for each of the following:

(a) the derivatives of sums, differences, products, and quotients;

(b) the derivatives of polynomials;

(c) the derivatives of e^z and e^{-z}.

4. Let $\phi(u)$ be a complex-valued function of the real variable u on the interval $[a,b]$, and suppose its real and

imaginary parts are given by differentiable functions $r(u)$ and $s(u)$, respectively, so that

$$\phi(u) = r(u) + is(u).$$

Then the derivative $\phi'(u)$ is defined by the equation

$$\phi'(u) = r'(u) + is'(u).$$

Similarly, the integral $\int_a^b \phi'(u)du$ is defined by the equation

$$\int_a^b \phi'(u)du = \int_a^b r'(u)du + i\int_a^b s'(u)du.$$

Prove the relevant form of the Fundamental Theorem of Calculus:

$$\phi(b) - \phi(a) = \int_a^b \phi'(u)du.$$

5. Let x be a fixed complex number, g a differentiable complex valued function of a complex variable, and u a real variable. Then the function $\phi(u) = g(ux)$ is a complex function of a real variable. Show that $\phi'(u) = g'(ux) \cdot x$.

6. In which part(s) of the proof of Theorem 10 was it necessary to use the fact that p is prime?

7. It was implicitly assumed in the proof of Theorem 10 that there are an infinite number of primes. Where was this assumed? Prove this fact.

8. Prove that the number e is transcendental. (Hint: This is quite a bit easier than the corresponding proof for π, but can be developed along the same outline. In particular, the assumption that e is algebraic leads immediately to an equation very similar to the equation $e^{q_1} + e^{q_2} + \cdots + e^{q_m} + k = 0$, but where the exponents are integers and there may be integral coefficients in front of the terms. Working backwards, the reader can

construct an analogue of g, and from this analogues of f and F. The proof proceeds almost as before, but without the complication of symmetric functions. However, some coefficients need to be inserted in summing. This problem should help the reader gain a better understanding of the proof for π.)

REFERENCES AND NOTES

An encyclopedic reference on construction problems is [1]. The author mentions (p. 153) that the common elementary method of solution which we have used in Sections 1.2 and 1.3 goes back to Edmund Landau, when he was a student. Elementary treatments in English of construction problems may be found in [2, 3]. These references all discuss further various constructibility questions when the methods of construction are modified. It can be shown, for example, that any number that is constructible by compass and straightedge can actually be constructed by compass alone. (Of course we can't draw segments then, but we can construct points the appropriate distance apart.) The proof we have given for the transcendence of π is based on [4]. The original paper of Lindemann is [5], and additional proofs may be found in [6, 7] and elsewhere. As mentioned in the Introduction, a fascinating history of false attempts at squaring the circle is contained in [8]. Proofs of the Fundamental Theorem of Algebra go back to the doctoral thesis of Gauss. For rather simple proofs of this see [2, 9]; alternatively one may consult standard complex variables texts such as [10, 11], the latter containing five different proofs. An excellent discussion of polynomials, symmetric functions, and algebraic numbers may be found in [12], which then moves in directions different from the present volume and which the reader may well enjoy perusing. Three basic references on countability questions are [2, 13, 14], in increasing order of detail and difficulty.

1. Ludwig Bieberbach, *Theorie der Geometrischen Konstruktionen*, Birkhäuser, Basel, 1952.

2. Richard Courant and Herbert Robbins, *What Is Mathematics?*, Oxford University Press, New York, 1941.

3. Felix Klein, *Famous Problems in Elementary Geometry*, tr. by W. Beman and D. Smith, Ginn, Boston, 1897; reprinted by Chelsea, New York, 1956.

4. Ivan Niven, The transcendence of π, *Amer. Math. Monthly*, 46 (1939) 469–471.

5. F. Lindemann, Über die Zahl π, *Math. Ann.*, 20 (1882) 213–225.

6. Nathan Jacobson, *Basic Algebra I*, Freeman, San Francisco, 1974.

7. R. Steinberg and R. M. Redheffer, Analytic proof of the Lindemann Theorem, *Pacific J. Math.*, 2 (1952) 231–242.

8. F. J. Duarte, *Monografía sobre los Números π y e*, Estados Unidos de Venezuela, *Bol. Acad. Ci. Fís. Mat. Nat.*, 11 (1948), No. 34–35, 1–252 (1949).

9. Frode Terkelson, The fundamental theorem of algebra, *Amer. Math. Monthly*, 83 (1976) 647.

10. Ruel V. Churchill and James W. Brown, *Complex Variables and Applications*, rev. ed., McGraw-Hill, New York, 1974.

11. Norman Levinson and Raymond M. Redheffer, *Complex Variables*, Holden-Day, San Francisco, 1970.

12. Harry Pollard and Harold G. Diamond, *The Theory of Algebraic Numbers*, 2nd ed., Carus Mathematical Monographs, Number Nine, Mathematical Association of America, 1975.

13. Garrett Birkhoff and Saunders Mac Lane, *A Survey of Modern Algebra*, 4th ed., Macmillan, New York, 1977.

14. Edwin Hewitt and Karl Stromberg, *Real and Abstract Analysis*, Springer-Verlag, New York, 1965.

CHAPTER **2**

FIELD EXTENSIONS

Section 2.1. Arithmetic of Polynomials

In the last chapter, we saw how certain classical geometric construction problems depended for their solution on the nature of the roots of certain polynomials. We shall take another look at these problems later in this chapter. In the next chapter, we shall look at another famous problem, namely, the problem of trying to find analogues of the quadratic formula for the roots of polynomials of degree greater than 2. Since all these investigations deal with polynomials, we need first to learn more about the 'arithmetic' of polynomials.

Given a field $\mathbf{F}$, we denote by $\mathbf{F}[x]$ the set of all polynomials over $\mathbf{F}$ in the variable x. As we shall see, this set has many properties very similar to those of the set $\mathbf{Z}$ of integers. First, $\mathbf{F}[x]$ is closed under the operations of addition, subtraction, and multiplication. But, like $\mathbf{Z}$, it is not closed under division; the quotient of two polynomials is not necessarily a polynomial. In place of division, we have 'division with remainder'. For example, in $\mathbf{Z}$, we cannot simply divide 7 by 3; but we can say "7 divided by 3 is 2 with remainder 1"; and we can write this, entirely within $\mathbf{Z}$, as: $7 = 3 \cdot 2 + 1$; 2 is called the quotient and 1 is called the remainder. In $\mathbf{F}[x]$ we have similar 'division with remainder', namely, the **division algorithm for polynomials**:

THEOREM 11. *Let f and g be polynomials in* $\mathbf{F}[x]$, $g \neq 0$. *Then there exist unique polynomials q and r, also in* $\mathbf{F}[x]$, *with either r identically* 0 *or the degree of r less than that of g, such that $f = g \cdot q + r$. (q is called the quotient of f divided by g, and r is called the remainder.)*

Proof. First, let us note that q and r are simply what would emerge if we divided f by g according to the usual process of long division of polynomials, usually treated in high school. For a rigorous proof, let us write

$$f(x) = a_n x^n + a_{n-1} x^{n-1} + \cdots + a_0,$$

$$g(x) = b_m x^m + b_{m-1} x^{m-1} + \cdots + b_0, \qquad b_m \neq 0.$$

Let m be arbitrary but fixed; we prove the result for all values of $n \geqslant 0$. For $0 \leqslant n < m$, we must have $q = 0$ (that is, the 0 polynomial) and so $r = f$. Now we show the result by induction on n, for $n \geqslant m$. For $n = m$, q must be the constant a_n / b_m, since r cannot contribute to the leading term. This gives $r = f - (a_n / b_m) g$, thereby determining it uniquely as an element of $\mathbf{F}[x]$. Assuming now that the result is true up through polynomials of degree $n - 1$, we show it for n. In particular, since $f - (a_n / b_m) x^{n-m} g$ is of degree at most $n - 1$, there are unique polynomials $\tilde{q}$ and r, $r = 0$ or $\deg r < \deg g$, such that

$$f - (a_n / b_m) x^{n-m} g = g \cdot \tilde{q} + r,$$

whence

$$f = g \cdot \left[(a_n / b_m) x^{n-m} + \tilde{q} \right] + r.$$

The bracketed term is an acceptable q. It is unique since its leading term must be $(a_n/b_m)x^{n-m}$, and from this, by the induction, the sum of all terms of lower degree, given by $\tilde{q}$, is uniquely determined. Since q is unique, so too is r. ∎

It is useful to observe that if, in the context of Theorem 11, $\tilde{\mathbf{F}} \subset \mathbf{F}$ is a smaller field containing all the coefficients of f and g, then the proof implies that both q and r actually belong to $\tilde{\mathbf{F}}[x]$, a subset of $\mathbf{F}[x]$. As a consequence, if f and g belong to both $\mathbf{F}_1[x]$ and $\mathbf{F}_2[x]$, where $\mathbf{F}_1$ and $\mathbf{F}_2$ are fields, the result of the division algorithm will be the same in both settings because the division actually takes place in $\tilde{\mathbf{F}}[x]$, where $\tilde{\mathbf{F}} = \mathbf{F}_1 \cap \mathbf{F}_2$. We refer to this fact by saying that the division of one polynomial by another is independent of the field over which we consider them.

We say that a polynomial g *divides* a polynomial f, written $g|f$, if the remainder upon dividing f by g is the 0 polynomial. For example, we write $(x^2-1)|(x^4-1)$ since $(x^4-1) = (x^2-1)(x^2+1)$. In this case, we call g a *divisor* of f, and the expression f/g is used for the quotient of f divided by g. It is a *proper divisor* if its degree is less than that of f and a *nontrivial divisor* if its degree is greater than 0. With these concepts of divisibility, we can now introduce the analogue in $\mathbf{F}[x]$ of prime numbers in $\mathbf{Z}$. A nonconstant polynomial $f \in \mathbf{F}[x]$ is called *irreducible in* $\mathbf{F}[x]$ or *irreducible over* $\mathbf{F}$ if it has no nontrivial proper divisors in $\mathbf{F}[x]$. To put it another way, f is irreducible over $\mathbf{F}$ if it cannot be written as the product of two lower degree polynomials over $\mathbf{F}$. As an example, x^2+1 is irreducible over $\mathbf{Q}$. But irreducibility depends closely on the field in question. The same polynomial is also irreducible over $\mathbf{R}$; but over either $\mathbf{Q}(i)$ or $\mathbf{C}$ it is reducible, factoring into $(x - i)(x + i)$. In fact, over **C** every polynomial is *completely reducible*, that is, factorable into all linear factors. Given a polynomial over a field **F**, it is in general quite difficult to determine whether it is irreducible or not. For the field **Q**, a useful criterion will be given below in Theorem 12; and, fortunately, this is the case which holds the greatest importance for us. For the proof of this theorem, we shall make use of the following **Lemma of Gauss**:

LEMMA 12. *If a polynomial with integral coefficients can be written as the product of two lower degree polynomials with rational coefficients, then, in fact, it can be written as the product of two lower degree polynomials with integral coefficients.*

Proof. Suppose that $f = gh$, where f has integral coefficients and g and h have rational coefficients. (Here and elsewhere, juxtaposition will be understood to mean multiplication of polynomials.) Without loss of generality, we may assume that the coefficients of f are relatively prime (i.e., they have no common factor other than ± 1), for otherwise we may divide out the common factors, apply the result, and then multiply through again. Let us write

$$g(x) = \frac{a_n}{b_n} x^n + \frac{a_{n-1}}{b_{n-1}} x^{n-1} + \cdots + \frac{a_0}{b_0},$$

$$h(x) = \frac{c_m}{d_m} x^m + \frac{c_{m-1}}{d_{m-1}} x^{m-1} + \cdots + \frac{c_0}{d_0},$$

where the a's, b's, c's, and d's are integers. Multiplying the first equation by the product B of all the b's gives a polynomial on the right with integral coefficients, out of which we can factor their greatest common divisor A. Thus

$$(B/A)g(x) = A_n x^n + A_{n-1}x^{n-1} + \cdots + A_0$$

is a polynomial with relatively prime integral coefficients. In like manner, we obtain another such polynomial

$$(D/C)h(x) = C_m x^m + C_{m-1}x^{m-1} + \cdots + C_0.$$

And from $f = gh$ we obtain $BDf = (AC)[(B/A)g] \cdot [(D/C)h]$. Since BD is the *greatest* common divisor of the coefficients of this polynomial, as we see from looking at the left side of the equation, and since AC is a common divisor of the coefficients, as we see from the right, then $AC \mid BD$. Defining $E = BD/AC$, an integer, we have

$$\begin{aligned} Ef(x) &= \left[A_n x^n + A_{n-1}x^{n-1} + \cdots + A_0\right] \\ &\qquad \times \left[C_m x^m + C_{m-1}x^{m-1} + \cdots + C_0\right]. \end{aligned}$$

If we can show $E = \pm 1$, we shall be done; for we shall then have written f in the required form. If $E \neq \pm 1$, it has a prime factor p. But p cannot divide all the A_i's, for they are relatively prime, and so there exists a smallest integer $i \geqslant 0$ such that p does not divide A_i. Similarly, there is a smallest integer $j \geqslant 0$ such that p does not divide C_j. Now we compare on both sides of the equation the coefficient of x^{i+j}. From the left side, this coefficient is divisible by E and hence by p. From the right side, it is calculated to be

$$\sum_{k=0}^{i-1} A_k C_{i+j-k} + \sum_{k=i+1}^{i+j} A_k C_{i+j-k} + A_i C_j.$$

Since each A_k in the first sum is divisible by p, so too is that sum. Since each C_{i+j-k} in the second sum is divisible by p, so too is that sum. But by the definition of i and j, the term A_iC_j is not divisible by p. Hence the entire sum cannot be divisible by p, which is a contradiction. Hence $E = \pm 1$. ∎

We are now able to derive a sufficient condition for irreducibility over $\mathbf{Q}$ of polynomials with integral coefficients. It is called the **Eisenstein irreducibility criterion.**

THEOREM 12. *Let* $f(x) = a_nx^n + a_{n-1}x^{n-1} + \cdots + a_0$, *where the coefficients are all integers. If there exists a prime p such that*: (i) *p divides each of* $a_0, a_1, \ldots, a_{n-1}$; (ii) *p does not divide* a_n; *and* (iii) p^2 *does not divide* a_0; *then f is irreducible over* $\mathbf{Q}$.

Proof. If f were reducible, it could be written as a product gh of polynomials over $\mathbf{Q}$ each with degree $\geqslant 1$. By Lemma 12, we may assume g and h have integral coefficients. If we write

$$g(x) = b_kx^k + b_{k-1}x^{k-1} + \cdots + b_0,$$

$$h(x) = c_mx^m + c_{m-1}x^{m-1} + \cdots + c_0,$$

then $a_n = b_kc_m$, so that p cannot divide either b_k or c_m. Also, $a_0 = b_0c_0$, so p divides exactly one of b_0 and c_0; without loss of generality assume that p divides b_0 but that p does not divide c_0. Since p does not divide b_k, there must be a smallest subscript $j \geqslant 1$ such that p does not divide b_j. Now, in the equation $f = gh$ compare on both sides the coefficients of x^j. On the left it is a_j, which is divisible by p since $j \leqslant k < n$. On the right it is the sum $b_0c_j + b_1c_{j-1} + \cdots + b_jc_0$, which is not divisible by p because every term except the last is divisible by p. This gives a contradiction. Thus f cannot be reducible. ∎

It is important to realize that the above theorem only gives a sufficient condition for irreducibility over **Q**. For example, it enables us to conclude that $x^5 - 2$ and $3x^5 + 7x^4 - 14x^2 + 7x + 56$ are irreducible over **Q** (take $p = 2$ and 7 respectively), but it does not give us any immediate information about $x^3 - 3x - 1$ or $x^4 + x^3 + x^2 + x + 1$, both of which are also irreducible over **Q**. To treat situations like these, sometimes a simple change of variables can be found to put a polynomial in a form to which the theorem is applicable. For example, making the change of variables $x = u + 1$ in $x^3 - 3x - 1$, we obtain $u^3 + 3u^2 - 3$, which is now seen to be irreducible ($p = 3$). Clearly, the property of irreducibility is not changed by this transformation. Use will be made of this technique in the problems as well as in later sections.

So far, irreducible polynomials have simply been defined, and for the special case of **Q**$[x]$ a sufficient condition for irreducibility has been given. Noting that the definition of an irreducible polynomial is quite similar to that of a prime number, we shall now proceed to see that the analogy extends much further. One fact about primes which we have used all along is that if a prime divides a product of two numbers, it must divide one of them. A similar statement will turn out to be true for irreducible polynomials (Theorem 14). Another useful fact about primes is that every integer can be written uniquely as a product of primes. Almost the same statement will turn out to be true for the factorization of a polynomial into irreducible factors (Theorem 15). In order to develop these results, it is first necessary to introduce one more concept.

Given two polynomials f and g in $\mathbf{F}[x]$, a *greatest common divisor of f and g*, denoted (f, g), is any polynomial of maximal degree and in $\mathbf{F}[x]$ which divides both f and g. That is, there may be many polynomials (certainly including all the constant polynomials) in $\mathbf{F}[x]$ which divide both f and g; any one of maximal degree is a greatest common divisor (abbreviated g.c.d.) of f and g. Actually, it will turn out that all the g.c.d.'s of f and g are constant multiples of each other. When we write (f, g), we shall mean some or any g.c.d., as the context will make clear. Given f and g, how do we find (f, g)? Does it depend on the field $\mathbf{F}$? (Surprisingly, the answer to this second question is "no".) To begin on the first question, we first take a look at $\mathbf{Z}$.

Perhaps the reader has studied at some time the Euclidean algorithm, which is a procedure for finding the g.c.d. of two integers. It may be best to illustrate it by an example. To find, say, the g.c.d. of 882 and 270, first divide one by the other, as in the division algorithm:

$$882 = 270 \cdot 3 + 72.$$

Now divide the most recent divisor (270) by the most recent remainder (72):

$$270 = 72 \cdot 3 + 54.$$

Keep repeating this process until you get remainder 0:

$$72 = 54 \cdot 1 + 18$$

$$54 = 18 \cdot 3 + 0.$$

The divisor in the last step (18) is the g.c.d. Furthermore, by retracing the steps, we find that the g.c.d. can be written as a linear combination, with integral coefficients, of the original numbers:

$$\begin{aligned} 18 &= 72 - 54 \\ &= 72 - [270 - 72 \cdot 3] \\ &= [882 - 270 \cdot 3] - [270 - (882 - 270 \cdot 3) \cdot 3] \\ &= 4 \cdot 882 - 13 \cdot 270. \end{aligned}$$

This is an important fact, frequently more useful than the procedure itself. The reader is asked in Problem 2 to prove the validity of the Euclidean algorithm for **Z**. We shall now state and prove the corresponding **Euclidean algorithm for F**$[x]$, the actual procedure, which is not important for our purposes, being a by-product of the proof.

THEOREM 13. *Let* $f, g \in \mathbf{F}[x]$. *Then there exist also* s, $t \in \mathbf{F}[x]$ *such that* $(f, g) = sf + tg$.

Proof. Begin by dividing f by g, as in the division algorithm, to get

$$f = g \cdot q_1 + r_1.$$

Now divide the most recent divisor (g) by the most recent remainder (r_1):

$$g = r_1 \cdot q_2 + r_2.$$

Keep repeating the process. Eventually, some $r_{N+1} = 0$. (Make sure you see why.) Thus we have:

$$\begin{aligned} r_1 &= r_2 \cdot q_3 + r_3 \\ r_2 &= r_3 \cdot q_4 + r_4 \\ &\vdots \\ r_{N-2} &= r_{N-1} \cdot q_N + r_N \\ r_{N-1} &= r_N \cdot q_{N+1} + 0. \end{aligned}$$

It is easy to see that r_N can be written in the required form, for each r_i can be so written. To prove this, we work by induction on i. For $i = 1$, we have $r_1 = 1 \cdot f + (-q_1) \cdot g$. For $i = 2$, we have $r_2 = g - r_1 \cdot q_2 = g - f \cdot q_2 + q_1 q_2 \cdot g = (-q_2) \cdot f + (1 + q_1 q_2) \cdot g$. Assuming it true up to $i - 1$, we write $r_i = r_{i-2} - r_{i-1} \cdot q_i$ from which the desired conclusion is obvious. It remains to show that r_N is actually a g.c.d. From the representation $r_N = sf + tg$, we see that any common divisor of f and g, and hence any greatest common divisor of f and g, say (f, g), divides r_N, and so $\deg (f, g) \leqslant \deg r_N$. But by the equations leading to r_N, working backwards from the last one, we have r_N successively dividing $r_{N-1}, r_{N-2}, \ldots, r_1, g$, and f. Thus, r_N is a common divisor of f and g, so that $\deg r_N \leqslant \deg (f, g)$. Combining this with the previous result, we see that $\deg r_N = \deg (f, g)$, and thus r_N is a g.c.d. of f and g. Further, since any g.c.d. has been shown to divide r_N, it follows that all g.c.d.'s of f and g are constant multiples of r_N and hence of each other. Thus, the representation given in the theorem is possible for every g.c.d. of f and g. ∎

An immediate consequence of Theorem 13 is the fact that every common divisor of f and g divides (f, g), which property is frequently taken as the very definition of the g.c.d. The reader is asked to show in Problem 5 that the g.c.d. of two polynomials is independent of the field over which we consider them.

Following the terminology of **Z**, we say that two polynomials f and g are *relatively prime* if they have no common nonconstant factors. In this case we can write $(f, g) = 1$, where the right side naturally refers to the constant polynomial with value 1. Now at last we can prove the two basic results mentioned earlier.

THEOREM 14. *Let $f, g, h \in \mathbf{F}[x]$ and let f be irreducible. If $f \mid gh$, then $f \mid g$ or $f \mid h$.*

Proof. Suppose f does not divide g. Then $(f, g) = 1$ by the fact that f has no nonconstant factors other than itself or constant multiples of itself. By Theorem 13, there exist polynomials s and $t \in \mathbf{F}[x]$ such that

$$1 = sf + tg,$$

so that

$$h = sfh + tgh,$$

from which, since $f \mid gh$, we conclude that $f \mid h$. ∎

THEOREM 15. *Let $f \in \mathbf{F}[x]$. Then f can be written as a product of irreducible polynomials in $\mathbf{F}[x]$, and this decomposition is unique up to constant multiples of each nonconstant factor.*

Proof. Of course, the result is trivial for f a constant polynomial. When $\deg f \geqslant 1$, it is obvious that f can be written as a product of nonconstant irreducible factors: just keep factoring it until each factor is irreducible. To show essential uniqueness, suppose that we can simultaneously write $f = p_1 p_2 \cdots p_n = q_1 q_2 \cdots q_m$, where the p's and q's are nonconstant irreducible polynomials in $\mathbf{F}[x]$. By Theorem 14, p_1 must divide some q_i, so by the latter's irreducibility q_i is a constant multiple of p_1. When both sides of the equation are divided by p_1, the results must be equal, by the uniqueness assertion of the division algorithm. We now repeat the process for p_2, and then p_3, and so on. This also implies that $n = m$; for otherwise we would reach a point where one side has degree 0 and the other side does not. ∎

PROBLEMS

1. Prove the division algorithm for $\mathbf{Z}$: If $m, n \in \mathbf{Z}$, $n \neq 0$, then there exist unique integers q and r, with $0 \leqslant r < |n|$, such that $m = nq + r$.

2. Prove the Euclidean algorithm for $\mathbf{Z}$: If $m, n \in \mathbf{Z}$, $n \neq 0$, then there exist integers s and t such that the g.c.d. $(m, n) = sm + tn$. Moreover, the g.c.d. can be obtained in a finite number of steps by the method of repeated division of the most recent divisor by the most recent remainder, beginning with the division of m by n according to the division algorithm, as outlined in the text just prior to Theorem 13.

3. Find the g.c.d. of 264 and 714. Can you take them in either order?

4. Use the Euclidean algorithm for $\mathbf{Q}[x]$ to find a g.c.d. of $f(x) = 2x^6 - 10x^5 + 2x^4 - 7x^3 - 15x^2 + 3x - 15$ and $g(x) = x^5 - 4x^4 - 3x^3 - 9x^2 - 4x - 5$.

5. Show that the g.c.d. of two polynomials is independent of the field over which we consider them. That is, suppose $\mathbf{F}_1$ and $\mathbf{F}_2$ are fields and f and g are polynomials belonging to both $\mathbf{F}_1[x]$ and $\mathbf{F}_2[x]$; show that they have the same g.c.d. in both $\mathbf{F}_1[x]$ and $\mathbf{F}_2[x]$. (Note: Some divisibility properties, such as irreducibility, are closely related to the field in question, but the g.c.d. is seen here not to be.)

6. Let $f, g \in \mathbf{F}[x]$, f irreducible. Suppose there exists some $a \in \mathbf{C}$ such that $f(a) = g(a) = 0$; that is, they have a common root in $\mathbf{C}$. Show that f divides g.

7. Let $f, g, h \in \mathbf{F}[x]$, $f \neq 0$ (i.e., to recall our convention, f is not the 0 polynomial). If $fg = fh$, show that $g = h$.

8. Show that $(x^n - 1)/(x - 1)$ is a polynomial over $\mathbf{Q}$ for all positive integers n, and determine precisely the set of values of n for which it is irreducible.

9. Decompose $x^{10} - 1$ into a product of irreducible factors in $\mathbf{Q}[x]$.

10. Let f be a polynomial irreducible over $\mathbf{F}$. Show that f has no multiple roots. (Hint: Consider the g.c.d. of f and its derivative.)

11. A polynomial with integral coefficients is said to be *primitive* if its coefficients are relatively prime (that is, there is no integer other than ± 1 which divides all the coefficients). If f is any polynomial with integral coefficients and if g is a primitive polynomial which divides f, show that the quotient f/g actually has integral coefficients. (As an application of this, observe that if a monic polynomial with integral coefficients divides another polynomial with integral coefficients, then the quotient has integral coefficients.)

12. If f is a polynomial with integral coefficients and if g is a monic polynomial with integral coefficients, show that the quotient q and the remainder r, resulting from the division of f by g according to the division algorithm, both have integral coefficients.

*13. Develop an algorithm for factoring any polynomial in $\mathbf{Q}[x]$ into irreducible factors.

Section 2.2. Simple, Multiple, and Finite Extensions

The matter of extensions of fields arose as early as Section 1.1, in which it was shown how quadratic extensions are fundamental to the theory of constructibility. In this section we shall generalize the notion of field extensions and develop a coherent theory within which we shall then be able to go back and look again at some earlier results.

If $\mathbf{F}$ is a field and if $a_1, a_2, \ldots, a_m$ are any m complex numbers, we define the *extension of* $\mathbf{F}$ *by* $a_1, a_2, \ldots, a_m$, denoted $\mathbf{F}(a_1, a_2, \ldots, a_m)$, to be the smallest field containing $a_1, a_2, \ldots, a_m$ as well as all the elements of **F**. (By "smallest", of course, we mean the field which is the intersection of all fields containing all these elements.) In general, it is called a *multiple extension* of **F**, and in the particular case $m = 1$, a *simple extension.* In addition, if each of the a_i's is algebraic over **F**, it is called an *algebraic extension*; otherwise it is a *transcendental extension*.

It is natural to ask "What does $\mathbf{F}(a)$ look like?". That is, when a is adjoined to **F**, what other elements must also be adjoined so that we still have a field? Questions of this nature were first pursued in Section 1.1, especially in the problems at the end of the section. For example, we found that $\mathbf{Q}(\sqrt{2}) = \{a + b\sqrt{2} \mid a, b \in \mathbf{Q}\}$, $\mathbf{Q}(\sqrt{2}, \sqrt{3}) = \{a + b\sqrt{2} + c\sqrt{3} + d\sqrt{6} \mid a, b, c, d \in \mathbf{Q}\}$, and $\mathbf{Q}(\sqrt[5]{2}) = \{a + b2^{1/5} + c2^{2/5} + d2^{3/5} + e2^{4/5} | a, b, c, d, e \in \mathbf{Q}\}$. In these cases, then, we were able to express each element of the extension as a linear combination of some specific finite collection of elements in the extension, the coefficients in the linear combination coming from the original field. This is strongly suggestive of some vector space structure, which is precisely the key to studying these extensions.

First, to generalize even further the concept of field extensions, if **E** and **F** are fields such that $\mathbf{E} \supset \mathbf{F}$, then **E** is called an *extension* of **F**. In this case, **E** may be considered as a vector space over **F**, where vector addition is simply the usual field addition and scalar multiplication is the usual field multiplication. (The reader who is not accustomed to thinking of a field as a vector space over itself, or of the closely related notion here, may wish to work through Problem 1 before proceeding further.) The dimension of the vector space **E** over **F** is called the *degree* of the extension, and it is denoted $[\mathbf{E} : \mathbf{F}]$.

For example, $[\mathbf{Q}(\sqrt{2}) : \mathbf{Q}] = 2$, since a basis for $\mathbf{Q}(\sqrt{2})$ over **Q** is given by the set $\{1, \sqrt{2}\}$. Similarly,

$[\mathbf{Q}(\sqrt{2}, \sqrt{3}) : \mathbf{Q}] = 4$, since this extension has as a basis $\{1, \sqrt{2}, \sqrt{3}, \sqrt{6}\}$. That this set is a spanning set for the extension was shown in Problem 8 of Section 1.1; that it is a linearly independent set can be established by the type of argument used in Problem 7 of Section 1.1 and in Problem 2 of Section 1.4. In the case of $[\mathbf{Q}(\sqrt[5]{2}) : \mathbf{Q}]$ we know it is at most 5, since $\{1, 2^{1/5}, 2^{2/5}, 2^{3/5}, 2^{4/5}\}$ were shown in Problem 6 of Section 1.1 to form a spanning set for the extension, which itself took a good deal of work; the question of linear independence was not treated. Recalling these labors, it is natural to ask: **Is there some simple way to determine the degree of an extension?** The affirmative answer to this question in certain important cases, of which we shall make frequent use, will be developed in the next few paragraphs.

Let us begin with the case of a simple algebraic extension $\mathbf{F}(a)$ over $\mathbf{F}$. Since a is algebraic over $\mathbf{F}$, there exists a polynomial over $\mathbf{F}$ having a as a root; and hence there exists such a polynomial of smallest degree. Any such polynomial is called a *minimal polynomial* for a over $\mathbf{F}$, and its degree is called the *degree of a over* $\mathbf{F}$, denoted $\deg_{\mathbf{F}} a$. For example, $\deg_{\mathbf{R}} i = 2$, since $x^2 + 1$ is a minimal polynomial for i over $\mathbf{R}$. Given a number a, it is in general quite difficult to find directly (or to prove you have found) a minimal polynomial for a. The key to the matter is the equivalence between the concept of "minimal polynomial" and that of "irreducible polynomial":

THEOREM 16. *Let $f \in \mathbf{F}[x]$ and suppose $f(a) = 0$. Then f is a minimal polynomial for a over* $\mathbf{F}$ *if and only if f is irreducible over* $\mathbf{F}$.

Proof. If f is minimal, then it must certainly be irreducible, for if it could be written as a product gh of polynomials in $\mathbf{F}[x]$ of lower degree, one of them would have to have a as a root, contrary to the assumption of minimality.

On the other hand, if f is irreducible, we may apply the division algorithm to divide it by some minimal polynomial for a, say g. Thus $f = g \cdot q + r$. Since $f(a) = g(a) = 0$, we have $r(a) = 0$, so by the minimality of g, r must be the 0 polynomial. But, since f is irreducible and $\deg g \geqslant 1$, we must have $\deg g = \deg f$, so that q is a constant. Hence f, since it has the same degree as g, is also minimal. ∎

Since we have an irreducibility criterion for polynomials in $\mathbf{Q}[x]$ with integral coefficients (Theorem 12), we are sometimes able to use this theorem to find the degree of an algebraic number over $\mathbf{Q}$. For example, $\deg_{\mathbf{Q}} \sqrt[5]{2} = 5$, since $\sqrt[5]{2}$ is a root of the irreducible polynomial $x^5 - 2$. Also, $\deg_{\mathbf{Q}} \cos 20° = 3$, since $\cos 20°$ is a root of the irreducible polynomial $8x^3 - 6x - 1$ (see Section 1.3).

With these ideas in mind, we now return to the problem of finding the degree of a simple algebraic extension:

THEOREM 17. *If a is algebraic over* $\mathbf{F}$, $[\mathbf{F}(a) : \mathbf{F}] = \deg_{\mathbf{F}} a$. *If a is transcendental over* $\mathbf{F}$, $[\mathbf{F}(a) : \mathbf{F}] = \infty$.

Proof. For any a, we observe that

$$\mathbf{F}(a) = \{ g(a)/h(a) \mid g,h \in \mathbf{F}[x], h(a) \neq 0\},$$

for this latter set is first of all a field (it satisfies the definition) and all its elements must be contained in any field containing a and all of $\mathbf{F}$.

Now, when a is algebraic over $\mathbf{F}$ we are able to simplify the representation of the elements in this set. In particular, letting $\deg_{\mathbf{F}} a = n$, we first show that the elements

$1, a, a^2, \ldots, a^{n-1}$ are a spanning set for $\mathbf{F}(a)$ over $\mathbf{F}$; or what is equivalent, every element $g(a)/h(a)$ equals some $r(a)$, where $r \in \mathbf{F}[x]$ and $\deg r \leqslant n-1$ or $r = 0$. Let f be a minimal polynomial for a. If $\deg h \geqslant 1$, then $(f, h) = 1$, for f cannot divide h (or else $h(a) = 0$), and f is irreducible (by Theorem 16). By the Euclidean algorithm there exist $s, t \in \mathbf{F}[x]$ such that $1 = sf + th$. Substituting the value a for x and noting that $f(a) = 0$, we obtain $t(a) = 1/h(a)$. So now we can write $g(a)/h(a) = g(a)t(a)$. Application of the division algorithm to divide the polynomial gt by f yields $gt = fq + r$, where either $\deg r \leqslant n-1$ or $r = 0$. Again substituting the value a for x and recalling that $f(a) = 0$, we have $g(a)t(a) = r(a)$.

But the elements $1, a, a^2, \ldots, a^{n-1}$ are also linearly independent over $\mathbf{F}$, or else a would satisfy a polynomial over $\mathbf{F}$ of degree $\leqslant n-1$, contrary to the minimality of n. Hence these elements form a basis for the extension, completing the proof that when a is algebraic, $[\mathbf{F}(a) : \mathbf{F}] = \deg_{\mathbf{F}} a$.

When a is transcendental over $\mathbf{F}$, we can find linearly independent sets with arbitrarily many elements, namely $\{1, a, a^2, \ldots, a^n\}$ for any n. This set is linearly independent since a nontrivial linear combination of its elements would give us a polynomial f over $\mathbf{F}$ such that $f(a) = 0$. Hence $[\mathbf{F}(a) : \mathbf{F}]$ cannot be finite. ∎

If $\mathbf{E}$ is an extension of $\mathbf{F}$ and $[\mathbf{E} : \mathbf{F}]$ is finite, we call $\mathbf{E}$ a *finite extension* of $\mathbf{F}$. In a certain sense, the case of finding the degree of a finite or of a multiple algebraic extension of $\mathbf{F}$ is covered by the previous theorem, for we show below that these kinds of extensions are themselves simple extensions! But while this result is of interest in itself and important for later work, it does not provide us with a very useful way of finding or representing the degree of these extensions.

THEOREM 18. *The following statements are equivalent:*

(i) **E** *is finite extension of* **F**.

(ii) **E** *is a multiple algebraic extension of* **F**.

(iii) **E** *is a simple algebraic extension of* **F**.

Proof. (i⇒ii) We first observe that every element of **E** is algebraic over **F**. Let $a \in \mathbf{E}$ and define $n = [\mathbf{E} : \mathbf{F}]$. Then since the $n+1$ vectors $1, a, a^2, \ldots, a^n$ must be linearly dependent, there is a nontrivial linear combination of them equal to 0, with coefficients in **F**. Thus a is a root of some polynomial over **F**. Now, if we take a basis for **E** over **F**, say $a_1, a_2, \ldots, a_n$, it is easy to see that **E** is simply the multiple algebraic extension $\mathbf{F}(a_1, a_2, \ldots, a_n)$. For **E**, in that it contains all the elements of **F** as well as $a_1, a_2, \ldots, a_n$, must contain the *smallest* field containing these elements, which by definition is $\mathbf{F}(a_1, a_2, \ldots, a_n)$. Likewise, by closure under multiplication and addition, $\mathbf{F}(a_1, a_2, \ldots, a_n)$ must contain the set of all linear combinations of the a_i's, with coefficients in **F**, which set is precisely **E**.

(ii⇒iii) First we note that the multiple extension $\mathbf{F}(a_1, a_2, \ldots, a_n)$ can be achieved by a sequence of simple extensions: $\mathbf{F}_1 = \mathbf{F}(a_1), \mathbf{F}_2 = \mathbf{F}_1(a_2), \ldots, \mathbf{F}_n = \mathbf{F}_{n-1}(a_n)$. For $\mathbf{F}(a_1, a_2, \ldots, a_n)$ and $\mathbf{F}_n$ must each contain all the elements of the other. Thus, if we can show that these extensions can be combined two at a time to form simple extensions, we can repeatedly apply this result to deduce that $\mathbf{F}(a_1, a_2, \ldots, a_n)$ is a simple extension of **F**.

It suffices then to consider the case $n = 2$, and, altering the notation, to show that for any two numbers b and c that are algebraic over **F**, there exists a number a such that $\mathbf{F}(b, c) = \mathbf{F}(a)$. We shall actually find an acceptable a in the form of a simple linear combination of b and c, $a = b + kc$, where the coefficient k is a suitably chosen element of **F**.

Since b is algebraic over $\mathbf{F}$, it is a root of some minimal, and hence irreducible, polynomial f over $\mathbf{F}$, whose complete set of roots are $b = b_1, b_2, \ldots, b_q$. These roots are all distinct since an irreducible polynomial cannot have multiple roots (Problem 10 of Section 2.1). Similarly, c is the root of some irreducible polynomial g, whose roots are $c = c_1, c_2, \ldots, c_m$, where these are also all distinct. Look at the $q(m-1)$ equations in x: $b_i + xc_j = b_1 + xc_1$, for $1 \leqslant i \leqslant q$, $2 \leqslant j \leqslant m$. Since for these values of j, $c_j \neq c_1$, each one of these equations has exactly one solution in $\mathbf{C}$ for x, and hence at most one solution in $\mathbf{F}$ for x. Since $\mathbf{F}$ has an infinite number of elements, we can choose a number $k \in \mathbf{F}$ such that it is not any one of these solutions. Thus, $b_i + kc_j \neq b_1 + kc_1$ whenever $1 \leqslant i \leqslant q$ and $2 \leqslant j \leqslant m$.

Defining $a = b + kc$, we will now show that $\mathbf{F}(b, c) = \mathbf{F}(a)$. First, since $a \in \mathbf{F}(b, c)$, we have $\mathbf{F}(a) \subset \mathbf{F}(b, c)$. To show that $\mathbf{F}(b, c) \subset \mathbf{F}(a)$, it suffices to show that both b and $c \in \mathbf{F}(a)$. Indeed if we simply show that $c \in \mathbf{F}(a)$ we shall be done, for then $b = a - kc \in \mathbf{F}(a)$, since $\mathbf{F}(a)$ is a field.

We will show that $c \in \mathbf{F}(a)$ by showing that c is the root of a polynomial of degree 1 over $\mathbf{F}(a)$. Look at the polynomials $g(x)$ and $f(a - kx)$, both of which are certainly polynomials over $\mathbf{F}(a)$. (In fact, $g \in \mathbf{F}[x] \subset (\mathbf{F}(a))[x]$.) The number c is a root of each, as $g(c) = 0$ and $f(a - kc) = f(b) = 0$. Since their g.c.d. is divisible by $(x - c)$ in $\mathbf{C}[x]$, it has c as a root; the multiplicity of c as a root of the g.c.d. is exactly 1 since c is a root of multiplicity 1 of g. But, by the choice of k, the two polynomials $g(x)$ and $f(a - kx)$ have no other common roots, for the other roots of g are of the form c_j, $2 \leqslant j \leqslant m$, and $a - kc_j \neq b_i$ for any i. Thus, the g.c.d. of $g(x)$ and $f(a - kx)$ has degree 1, for all of its roots are common roots of these polynomials. But the g.c.d. is a polynomial over $\mathbf{F}(a)$, the field of coefficients of $g(x)$ and $f(a - kx)$. Thus c, a root of the g.c.d., is a root of a polynomial of degree 1 over $\mathbf{F}(a)$. Hence $c \in \mathbf{F}(a)$.

(iii $\Rightarrow$ i) This is immediate from Theorem 17. ∎

Much more important in the actual calculation of the degree of field extensions is the following theorem:

THEOREM 19. *Suppose* $\mathbf{E}$, $\mathbf{K}$, *and* $\mathbf{F}$ *are fields such that* $\mathbf{E} \supset \mathbf{K} \supset \mathbf{F}$. *Then* $[\mathbf{E} : \mathbf{F}] = [\mathbf{E} : \mathbf{K}][\mathbf{K} : \mathbf{F}]$, *where if either factor on the right is* ∞, *the product is taken to be* ∞.

Proof. Suppose $e_1, e_2, \ldots, e_n$ are linearly independent vectors of $\mathbf{E}$ over $\mathbf{K}$, and $k_1, k_2, \ldots, k_m$ are linearly independent vectors of $\mathbf{K}$ over $\mathbf{F}$. Then the nm vectors $e_i k_j$, $1 \leqslant i \leqslant n$, $1 \leqslant j \leqslant m$, are linearly independent vectors of $\mathbf{E}$ over $\mathbf{F}$. For, if we have a linear combination of them equal to 0, $\sum_{i=1}^{n} \sum_{j=1}^{m} (a_{ij} k_j) e_i = 0$, $a_{ij} \in \mathbf{F}$, then by the linear independence of the e_i's over $\mathbf{K}$, it follows that for each i, $\sum_{j=1}^{m} a_{ij} k_j = 0$. But now the linear independence of the k_j's over $\mathbf{F}$ implies that each $a_{ij} = 0$. This shows linear independence, since there can be no nontrivial linear combination equal to 0.

Thus, if either $[\mathbf{E} : \mathbf{K}]$ or $[\mathbf{K} : \mathbf{F}]$ is ∞, one of n or m may be taken arbitrarily large, implying that $[\mathbf{E} : \mathbf{F}] = \infty$. If they are both finite, we can take the e_i's and the k_j's in the previous paragraph as bases of their respective vector spaces. Thus it simply remains to show that the set of nm vectors $e_i k_j$ spans $\mathbf{E}$ over $\mathbf{F}$. Let v be an arbitrary element of $\mathbf{E}$. Since with $n = [\mathbf{E} : \mathbf{K}]$, the vectors, $e_1, e_2, \ldots, e_n$ form a basis for $\mathbf{E}$ over $\mathbf{K}$, there exist elements $b_1, b_2, \ldots, b_n$ of $\mathbf{K}$ such that $v = \sum_{i=1}^{n} b_i e_i$. But each b_i, being in $\mathbf{K}$, can be expressed in the form $b_i = \sum_{j=1}^{m} a_{ij} k_j$, where each $a_{ij} \in \mathbf{F}$. Combining this with the previous equation, $v=\sum_{i=1}^{n}\sum_{j=1}^{m}a_{ij}e_ik_j$, which represents v as a linear combination over $\mathbf{F}$ of the e_ik_j's. ∎

This relatively simple theorem will be applied often in the development of subsequent results. Let us observe here that by its repeated application to a finite sequence of field extensions $\mathbf{F}_N \supset \mathbf{F}_{N-1} \supset \cdots \supset \mathbf{F}_1 \supset \mathbf{F}_0$, we may conclude that

$$[\mathbf{F}_N : \mathbf{F}_0] = [\mathbf{F}_N : \mathbf{F}_{N-1}][\mathbf{F}_{N-1} : \mathbf{F}_{N-2}] \cdots [\mathbf{F}_1 : \mathbf{F}_0].$$

PROBLEMS

1. If $\mathbf{E}$ is an extension of $\mathbf{F}$, verify that $\mathbf{E}$ may be considered as a vector space over $\mathbf{F}$. (Recall the axioms for a vector space: A set $\mathbf{V}$ is said to be a *vector space over a field* $\mathbf{F}$ if it has the following properties: (a) there is a rule for adding vectors, always resulting in an element of $\mathbf{V}$; (b) this addition operation is associative and commutative; (c) there is a vector called 0, such that $v+0=v$ for every v; (d) for every v there is a vector called $-v$, such that $v+(-v)=0$; (*e*) there is a rule, called scalar multiplication, for multiplying a scalar times a vector, always resulting in an element of $\mathbf{V}$; (f) for $a, b \in \mathbf{F}$ and $u, v \in \mathbf{V}$, $(ab)v = a(bv)$, $a(u+v) = au + av$, and $(a+b)v = av + bv$; (g) for every $v \in \mathbf{V}$, $1v = v$, where 1 is simply the number 1 in $\mathbf{F}$.)

2. Show that if $\mathbf{E}$ is a quadratic extension of $\mathbf{F}$, then $[\mathbf{E} : \mathbf{F}] = 2$.

3. Let $\mathbf{E}$ be an extension of $\mathbf{F}$. Show that $[\mathbf{E} : \mathbf{F}] = 1$ if and only if $\mathbf{E} = \mathbf{F}$.

4. If $\mathbf{E}$ is an extension of $\mathbf{F}$ such that every element of $\mathbf{E}$ is algebraic over $\mathbf{F}$, we say that $\mathbf{E}$ is *algebraic* over $\mathbf{F}$. Is every algebraic extension a finite extension?

5. If $\mathbf{E}$ is an algebraic extension of $\mathbf{K}$ and $\mathbf{K}$ is an algebraic extension of $\mathbf{F}$, show that $\mathbf{E}$ is an algebraic extension of $\mathbf{F}$. (See the previous problem for the definition of an algebraic extension.)

6. Prove Theorem 8 by using the ideas of this section. (Hint: Observe that if a and b are algebraic over $\mathbf{F}$, then $\mathbf{F}(a, b)$ is a finite extension of $\mathbf{F}$.)

7. Express $\mathbf{Q}(\sqrt{2}, \sqrt{3})$ as a simple extension of $\mathbf{Q}$. Find the degree of this extension over $\mathbf{Q}$.

8. Let f be a polynomial of degree n over $\mathbf{F}$. Show that there exists an extension $\mathbf{E}$ of $\mathbf{F}$ in which f is completely reducible (i.e., can be factored into linear factors) and such that $[\mathbf{E} : \mathbf{F}] \leqslant n!$.

9. Let a be transcendental over $\mathbf{F}$. Show that the set $\{a^k \mid k \in \mathbf{Z}\}$ is linearly independent in $\mathbf{F}(a)$ over $\mathbf{F}$. Is it a basis for $\mathbf{F}(a)$ over $\mathbf{F}$?

10. Let $f \in \mathbf{R}[x]$ with $\deg f \geqslant 3$. Show that f is reducible.

11. Use the result of the previous problem to show that for $f \in \mathbf{R}[x]$, a is a root if and only if its complex conjugate $\bar{a}$ is a root. (The *complex conjugate* of $b + ci$ is $b - ci$, where $b, c \in \mathbf{R}$.)

*12. If $\mathbf{E}$ is a finite extension of $\mathbf{F}$, show that there are only a finite number of *intermediate fields* $\mathbf{K}$, $\mathbf{E} \supset \mathbf{K} \supset \mathbf{F}$. (This may be surprising.)

13. Suppose that $\mathbf{E} = \mathbf{F}(r)$, where r is a root of an irreducible polynomial f over $\mathbf{F}$. If $\tilde{r}$ is another root of f and if $\tilde{r} \in \mathbf{E}$, show that $\mathbf{E} = \mathbf{F}(\tilde{r})$.

Section 2.3. Geometric Constructions Revisited

A considerable amount of material was covered in the two previous sections, and so it is nice to be able to relax at this point and look back at some old, familiar construction problems, which have very simple solutions in terms of the recent material.

Recall that the duplication of the cube involved the construction of a root of the polynomial $x^3 - 2$, which is irreducible over $\mathbf{Q}$. Similarly, the trisection of an angle of 60° involved the construction of a root of the polynomial $x^3 - 3x - 1$, which is also irreducible over $\mathbf{Q}$ (Section 2.1). In each case it was required to construct a number whose degree over $\mathbf{Q}$ was 3. But by using the theory of field extensions, it is very simple to show that the only numbers that can possibly be constructed must have some power of 2 as their degree over $\mathbf{Q}$. That is, we have:

THEOREM 20. *If a is constructible, then* $\deg_{\mathbf{Q}} a$ *must be a power of* 2.

Proof. If a is constructible, then by Theorem 1 there is a sequence of fields $\mathbf{Q} = \mathbf{F}_0 \subset \mathbf{F}_1 \subset \cdots \subset \mathbf{F}_N$ such that $a \in \mathbf{F}_N$ and for each j, $\mathbf{F}_{j+1}$ is a quadratic extension of $\mathbf{F}_j$. Thus, for each j, $[\mathbf{F}_{j+1} : \mathbf{F}_j] = 2$, by Problem 2 of Section 2.2 (of course, it's practically obvious). By the repeated application of Theorem 19, $[\mathbf{F}_N : \mathbf{Q}] = 2^N$. Since $a \in \mathbf{F}_N$, $\mathbf{Q}(a) \subset \mathbf{F}_N$, so that by Theorem 19 again,

$$[\mathbf{F}_N : \mathbf{Q}] = [\mathbf{F}_N : \mathbf{Q}(a)][\mathbf{Q}(a) : \mathbf{Q}].$$

Since the left side is 2^N, neither factor on the right can have any odd prime factors. Thus, $[\mathbf{Q}(a) : \mathbf{Q}]$ must be a power of 2, and by Theorem 17 this equals $\deg_{\mathbf{Q}} a$. ∎

Let us even state separately, as a corollary, the following immediate reformulation:

COROLLARY 20. *It is not possible to construct any*

number which is the root of an irreducible polynomial over **Q** *that has degree other than a power of* 2.

It follows immediately from this result that it is not possible to duplicate the cube, as that would require the constructibility of a root of an irreducible polynomial over **Q** of degree 3. Similarly, it is not possible to trisect 60°, which shows that there can be no way (within the usual rules) to trisect an arbitrary angle. Although the original solutions to these problems were rather simple, one senses that the present method is more general and powerful. In fact, as the reader will see in Problem 1, it is now very easy to solve the problem of dividing an angle into five equal parts, which was quite difficult under the approach used in Chapter 1 for these other problems. (Cf. Problem 3 of Section 1.3.)

With these methods now available, we are finally able to tackle another famous problem: **For what values of *n* is it possible to construct a regular polygon of *n* sides**? Such a polygon is called a *regular n-gon*.

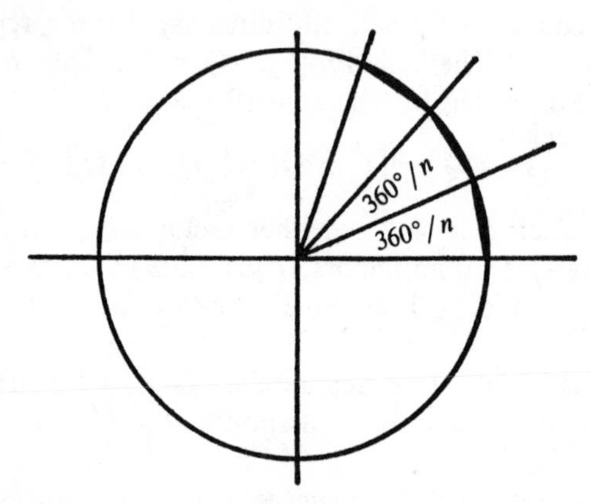

Figure 6

First we observe that the constructibility of a regular n-gon is equivalent to the constructibility of an angle of $360/n$ degrees. For, if $360/n$ degrees is constructible, we can inscribe a regular n-gon in the unit circle by successively marking off central angles of this amount, as illustrated in Figure 6. Conversely, if a regular n-gon is constructible, each of its exterior angles is constructible, and since these add up to $360°$, each is $360°/n$ (see Figure 7). Alternatively, one could construct the center of the circle in which the n-gon is inscribed by taking the intersection of the perpendicular bisectors of any two nonparallel sides. One then draws the circle, and the arcs between any two successive vertices each subtend a central angle of $360°/n$.

It would be possible at this point to find an equation relating $\cos(360°/n)$ and $\cos(360°)$, which equals 1 and hence is constructible, and to try to determine whether $\cos(360°/n)$ is constructible by using this equation. However, it is much simpler to effect a further reformulation of the problem first. The constructibility of an angle of $360°/n$ is obviously equivalent to the constructibility of each of the points $(\cos(k\,360°/n)$,

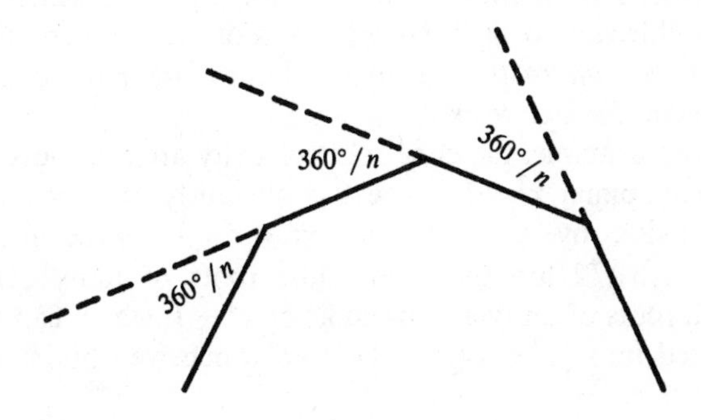

Figure 7

$\sin(k\,360°/n))$ in $\mathbf{R}^2$. In the next section we shall see that these points can be represented by complex numbers. Further, we shall see that the whole theory of constructibility can be extended rather simply to the complex numbers. Within this framework, we shall then be able to solve this problem.

PROBLEMS

1. Use the techniques of this section to show that it is not possible to divide an arbitrary angle into five equal parts by the usual rules of construction.

*2. Let $S = \{\theta \mid \theta \text{ is a trisectible angle}\}$. Prove that S is countable. (Keep in mind the distinction drawn at the end of Section 1.3.)

Section 2.4. Roots of Complex Numbers

Every complex number $a \neq 0$ has n distinct nth roots. For the polynomial $x^n - a$ has exactly n roots (Theorem 7), and none has multiplicity greater than one, for then it would also be a root of the derivative nx^{n-1}, which is impossible since $a \neq 0$. The nth roots of the number 1 are called the *nth roots of unity*, and they are particularly important for our work.

As an example, the cube roots of unity are the roots of the polynomial $x^3 - 1$. Since 1 is obviously one root, we can divide by $x - 1$ to get $x^2 + x + 1$ whose roots $(-1 \pm i\sqrt{3})/2$ are the other cube roots of unity. The fourth roots of unity are the roots of $x^4 - 1$, which can be factored into $(x^2 + 1)(x^2 - 1)$ from which we obtain the four values, $+i$, $-i$, $+1$, -1.

A convenient representation of the nth roots of unity is provided by the complex exponential function, introduced in Section 1.7. For by direct computation we find that the nth roots of unity are simply the numbers $e^{2\pi ik/n}$, where $1 \leqslant k \leqslant n$ and where $i = \sqrt{-1}$. To see this, we calculate

$$\left(e^{2\pi ik/n}\right)^n = e^{2\pi ik} = \left(e^{2\pi i}\right)^k = 1^k = 1,$$

and note that $e^{2\pi ik/n}$ gives n distinct numbers for $1 \leqslant k \leqslant n$. In terms of the trigonometric functions, $e^{2\pi ik/n} = \cos(2\pi k/n) + i\sin(2\pi k/n)$.

For a fixed n, some of the nth roots of unity may have the property that they are also mth roots of unity for some $m < n$. For example, although -1 is a fourth root of unity ($n = 4$), it is even a square root of unity ($m = 2$). Those nth roots of unity which are *not* mth roots of unity for any $m < n$ are called *primitive nth roots of unity*. There is always at least one primitive nth root of unity, as can be seen by taking $k = 1$ in the complex exponential form. If ω is *any* primitive nth root of unity, then the numbers $\omega, \omega^2, \omega^3, \ldots, \omega^n = 1$ are the entire list of nth roots of unity. To see this, observe first that they are all roots of unity, for $(\omega^j)^n = \omega^{jn} = (\omega^n)^j = 1^j = 1$. And they are distinct since if $\omega^j = \omega^k$ with $1 \leqslant k < j \leqslant n$, we would have $\omega^{j-k} = 1$, thus contradicting the primitive nature of ω, since $1 \leqslant j - k \leqslant n - 1$.

The *Euler ϕ-function*, defined by $\phi(n) =$ the number of integers between 1 and n inclusive which are relatively prime to n, provides a means of counting the number of primitive nth roots of unity, for we have:

THEOREM 21. *If ω is a primitive nth root of unity, then ω^k is also a primitive nth root of unity if and only if k and n are relatively prime. The number of primitive nth roots of unity is $\phi(n)$.*

Proof. The second sentence will clearly follow immediately from the first since all the roots of unity are represented by the values of k ranging from 1 to n. Now, suppose k and n are relatively prime. If there were some m, $1 \leqslant m < n$, such that $(\omega^k)^m = 1$, then km must be some integral multiple of n; otherwise, if $km = nq + r$, as in the division algorithm for $\mathbf{Z}$, $(\omega^k)^m = \omega^{km} = \omega^{nq}\omega^r = \omega^r = 1$, with $1 \leqslant r \leqslant n-1$, contrary to the primitive character of ω. Thus $n \mid km$; but since n and k have no common factors, $n \mid m$, which is impossible since $n > m$. Conversely, if k and n have a common prime factor p, then ω^k is an mth root of unity for $m = n/p < n$. For, $(\omega^k)^{n/p} = (\omega^n)^{k/p} = 1^{k/p} = 1$, since k/p is an integer. ∎

Let us also observe that if r is *any* nth root of the complex number $a \neq 0$ and if ω is a primitive nth root of unity, then the n distinct nth roots of a are given by $r\omega, r\omega^2, r\omega^3, \ldots, r\omega^n = r$.

There is a useful geometric interpretation of complex numbers and their arithmetical operations. Let $z = x + iy$, where x and y are real. We represent z in the Cartesian plane by the point (x, y). Similarly to polar coordinates we can write

$$\begin{aligned} z &= \sqrt{x^2+y^2}\left[\frac{x}{\sqrt{x^2+y^2}} + i\frac{y}{\sqrt{x^2+y^2}}\right] \\ &= |z|[\cos\theta + i\sin\theta] \\ &= |z|e^{i\theta}, \end{aligned}$$

since the point $(x/\sqrt{x^2+y^2}, y/\sqrt{x^2+y^2})$, being on the unit circle, can be represented by $(\cos\theta, \sin\theta)$ for some θ.

The angle θ is called an *argument* of z; the number $|z|=\sqrt{x^2+y^2}$ is called the *modulus* of z. The situation is depicted geometrically in Figure 8. The representation of a complex number z in the form $|z|e^{i\theta}$ is called its *polar representation*. It gives a very convenient way to picture the multiplication of complex numbers. Writing $z_1 = |z_1|e^{i\theta_1}$ and $z_2=|z_2|e^{i\theta_2}$, the laws of exponents imply that $z_1z_2=|z_1|\,|z_2|e^{i(\theta_1+\theta_2)}$, so that the modulus of a product is the product of the moduli and the argument of a product is the sum of the arguments. This is shown graphically in Figure 9. Thus, an nth root of a complex number z may be obtained by taking the real positive root of its modulus and dividing its argument by n. That is, an nth root of $z=|z|e^{i\theta}$ is given by $r=\sqrt[n]{|z|}\,e^{i\theta/n}$. It is important that the reader see this by *both* analytical *and*

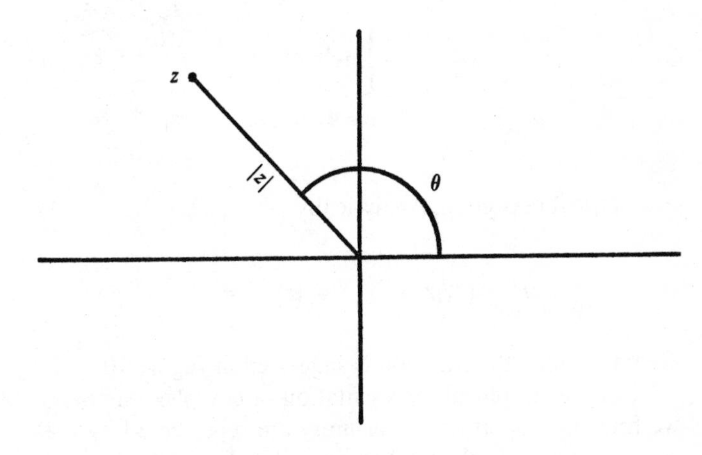

Figure 8

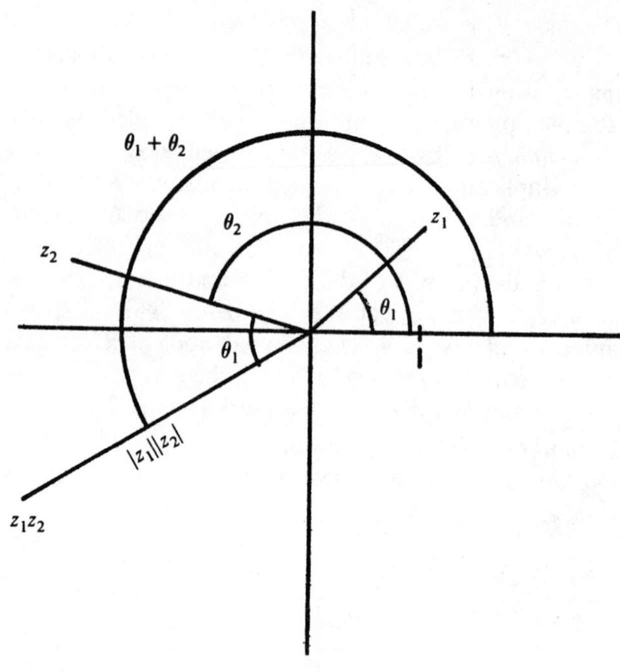

Figure 9

geometrical reasoning. Analytically,

$$r^n = \left(\sqrt[n]{|z|}\, e^{i\theta/n}\right)^n = |z|e^{i\theta} = z.$$

Geometrically, the situation is suggested in Figure 10.

From the graphical representation of complex numbers, we find that the nth roots of unity are represented by the points $(\cos(2\pi ik/n), \sin(2\pi ik/n))$, $1 \leqslant k \leqslant n$, which are

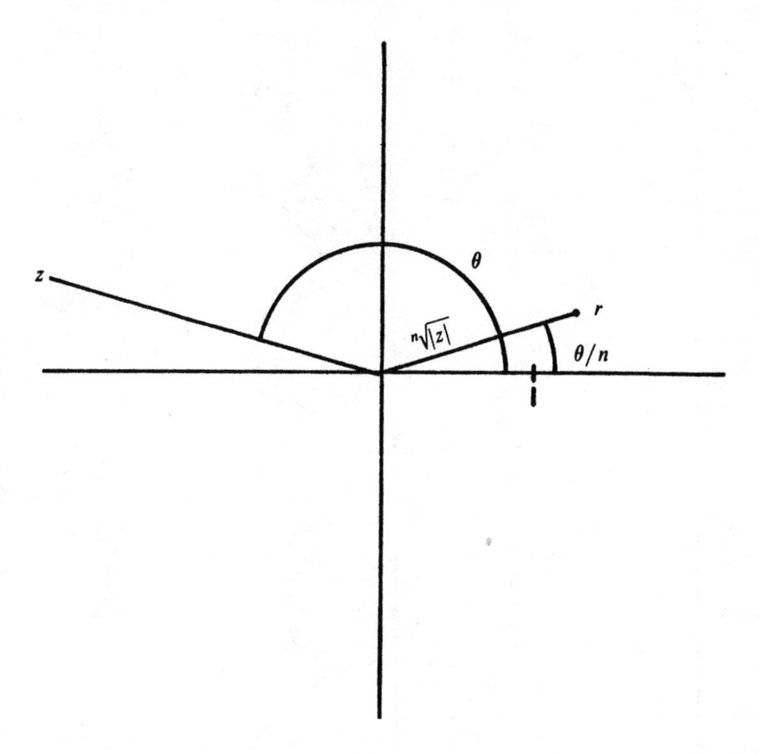

Figure 10

simply n equally spaced points around the unit circle, including the point $(1, 0)$. If we take $\omega = e^{2\pi i/n}$, these points are marked and labelled as powers of ω in Figure 11. Note that these are precisely the points that must be constructed in order to construct the regular n-gon!

With respect to the n nth roots of an arbitrary z, they are simply n equally spaced points on a circle of radius

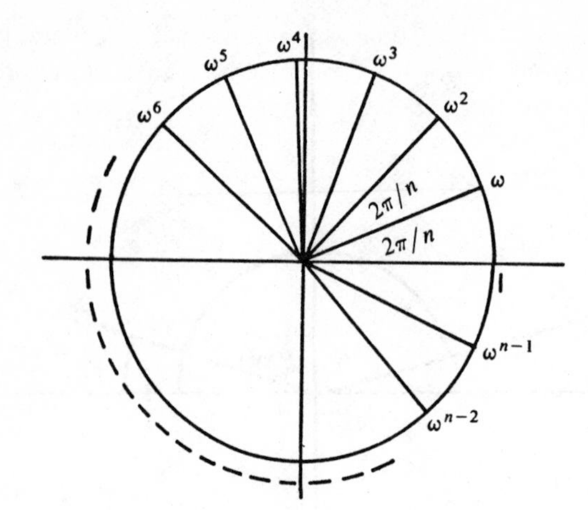

Figure 11

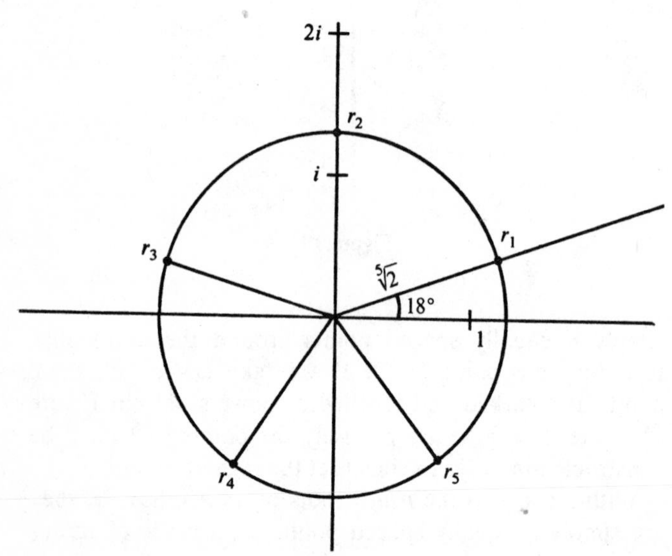

Figure 12

$\sqrt[n]{|z|}$; but, of course, their orientation will in general be different from that of the roots of unity. For example the five fifth roots of $2i$ are graphed in Figure 12.

PROBLEMS

1. Determine the cube roots of unity by a geometrical analysis.
2. Evaluate $\phi(k)$ for $k = 1, 2, \ldots, 15$.
3. Show that $\phi(p) = p - 1$ for p a prime.
4. Determine $\phi(p^n)$ for p a prime and n a positive integer.
5. Would it make sense to discuss "primitive" nth roots of 2?
6. How many primitive 20th roots of unity are there?
7. Describe the location of the seventh roots of $1 - i$ in the complex plane.
8. Show that if a is an nth root of unity, then a is a primitive mth root of unity for some m such that $m \mid n$.
9. Show that $\sum_{d|n}\phi(d) = n$, where the sum is over all positive integers d which divide n. Verify this result by direct calculation for the case $n = 12$.
10. Determine the degree over $\mathbf{Q}$ of the primitive nth roots of unity for $n = p$ and $n = p^2$, where p is a prime.
11. If k and n are relatively prime positive integers, show that $\sum_i r_i^k = 0$, where the r_i's are the n distinct nth roots of unity.

*12. Determine the degree over $\mathbf{Q}$ of each of the primitive 20th roots of unity. (This is not the same question as in Problem 6.)

*13. Show that the degree over **Q** of each of the primitive nth roots of unity when $n = p^m$, a power of a prime, is exactly $\phi(n)$.

Section 2.5. Constructibility of Regular Polygons I

In this section it will be possible to make substantial progress on the question formulated in Section 2.3, namely: **For what values of n is it possible to construct a regular n-gon?** It was seen in Section 2.4 that the constructibility of the regular n-gon is equivalent to the constructibility of the points in the Cartesian plane corresponding to the nth roots of unity. Since these are complex numbers, it will be convenient to extend the definition of constructibility directly to complex numbers.

For any number $a \in \mathbf{C}$, we say that a is *constructible* if both its real and imaginary parts are constructible. This is obviously equivalent to the constructibility of the point representing a in the Cartesian plane. The theory of constructibility extends immediately to complex numbers, as shown in Lemmas 22a and 22b and Theorems 22 and 23.

LEMMA 22a. *If a_1 and a_2 are constructible, then so too are $a_1 + a_2$, $a_1 - a_2$, $a_1 a_2$, and, when $a_2 \neq 0$, a_1/a_2.*

Proof. Let b_i and c_i denote, respectively, the real and imaginary parts of each a_i. From the computations below, we see that the real and imaginary parts of the results of combining a_1 and a_2 by each operation are each constructible by Lemma 1a, which is the analogue of the above for real numbers:

$$a_1 + a_2 = (b_1 + b_2) + i(c_1 + c_2)$$

$$a_1 - a_2 = (b_1 - b_2) + i(c_1 - c_2)$$

$$a_1 a_2 = (b_1 + ic_1)(b_2 + ic_2)$$

$$= (b_1 b_2 - c_1 c_2) + i(b_1 c_2 + b_2 c_1)$$

$$a_1/a_2 = (b_1 + ic_1)/(b_2 + ic_2)$$

$$= [(b_1 + ic_1)(b_2 - ic_2)]/[(b_2 + ic_2)(b_2 - ic_2)]$$

$$= \left[(b_1 b_2 + c_1 c_2)/(b_2^2 + c_2^2)\right]$$

$$+ i\left[(b_2 c_1 - b_1 c_2)/(b_2^2 + c_2^2)\right]. \qquad \blacksquare$$

LEMMA 22b. *If a is constructible and if $k^2 = a$, then k is constructible.*

Proof. It is obvious that a complex number $z = |z|e^{i\theta}$ is constructible if and only if the real number $|z|$ and the angle θ are constructible. Thus the number $|a|$ and an angle θ, an argument of a, are constructible. By Lemma 1b, $\sqrt{|a|}$ is constructible, and since an angle can be bisected, the angle $\theta/2$ is constructible. But $k = \pm\sqrt{|a|}\, e^{i\theta/2}$, so k is constructible. $\blacksquare$

The proof of the preceding lemma had a strong geometric flavor, and one should not hesitate to use such reasoning when it serves the purpose. The same kind of reasoning could be used to give simple alternate proofs of the multiplication and division parts of Lemma 22a. (See Problem 1.) Alternatively, Lemma 22b can be proved without reference to the polar form of a. (See Problem 2.)

In the statement of Lemma 22b, reference was made to a number k such that $k^2 = a$; but we did not write $\sqrt{a}$. The reason is this. Whereas for positive real numbers a, $\sqrt{a}$ is well defined as the unique positive square root of a, when a is complex, we have no such natural way to denote a particular square root of a. With this caution in mind, we shall agree to use the symbol $\sqrt{a}$ or $\sqrt[n]{a}$ as an abbreviation for an arbitrary square or nth root of a in cases where the statement we make is true for all such roots. Thus, it would now be acceptable to write Lemma 22b as: "If a is constructible, then $\sqrt{a}$ is constructible." To be consistent with this new convention, if a is a positive real number, we shall use the symbol $+\sqrt{a}$ to denote its positive square root.

By a *quadratic extension* of an arbitrary field $\mathbf{F}$ we mean a simple extension $\mathbf{F}(\sqrt{a})$, where $a \in \mathbf{F}$ and $\sqrt{a} \notin \mathbf{F}$; since the square roots of a are negatives of each other, the extension by either is the same field. Theorem 1 provided a necessary and sufficient condition for constructibility of real numbers, and it extends quite easily to the complex case:

THEOREM 22. *The following two statements are equivalent*:

(i) *The number a is constructible.*

(ii) *There exists a finite sequence of fields $\mathbf{Q} = \mathbf{F}_0 \subset \mathbf{F}_1 \subset \cdots \subset \mathbf{F}_N$, such that $a \in \mathbf{F}_N$ and for each j, $0 \leqslant j \leqslant N-1$, $\mathbf{F}_{j+1}$ is a quadratic extension of $\mathbf{F}_j$.*

Proof. (i$\Rightarrow$ii) Suppose $a = b + ic$, $b, c \in \mathbf{R}$. By Theorem 1 and the constructibility of b and c, there are two sequences of quadratic extensions: $\mathbf{Q} = \mathbf{G}_0 \subset \mathbf{G}_1 \subset$

$\cdots \subset \mathbf{G}_L$ with $b \in \mathbf{G}_L$ and $\mathbf{Q} = \mathbf{H}_0 \subset \mathbf{H}_1 \subset \cdots \subset \mathbf{H}_M$ with $c \in \mathbf{H}_M$. For each j, $\mathbf{H}_{j+1} = \mathbf{H}_j(\sqrt{k_j}\,)$. If we extend $\mathbf{G}_L$, one step at a time, by $\sqrt{k_0}\,, \sqrt{k_1}\,, \ldots, \sqrt{k_{m-1}}\,$, and i, omitting in the sequence of fields any repetitions, we have the desired sequence of quadratic extensions, for the final field contains b, c, and i, and hence a.

(ii$\Rightarrow$i) By induction on N. If $N = 0$, a is rational and hence constructible. Now we assume the theorem is true for $N = R$ and we prove it for $N = R + 1$. If $a \in \mathbf{F}_{R+1}$, then $a = a_R + b_R\sqrt{k_R}\,$, where a_R, b_R, and k_R all belong to $\mathbf{F}_R$, and hence are constructible. But then, by Lemmas 22a and 22b, a is constructible. ∎

By exactly the same reasoning as before, we use Theorem 22 to obtain the analogues for complex numbers of Theorem 20 and its Corollary:

THEOREM 23. *If a is constructible, then* $\deg_{\mathbf{Q}} a$ *must be a power of* 2. *Thus it is not possible to construct any number which is the root of an irreducible polynomial over* $\mathbf{Q}$ *that has degree other than a power of* 2.

Now we are ready to tackle the question stated at the beginning of this section. The constructibility of the regular n-gon is equivalent to the constructibility of the nth roots of unity, and hence, in view of Lemma 22a, to the constructibility of any primitive nth root of unity (since any nth root of unity is a product of a primitive root with itself an appropriate number of times). Our approach will be in two parts. First we will use Theorem 23 to narrow down the candidates for values of n for which the regular n-gon might be constructible. Then we shall show directly that for each one of these candidates the construction is actually possible.

One way to proceed with the first task would be to compute the degree of a primitive nth root of unity, and then to rule out all those values of n for which this degree is not a power of 2. It turns out that the $\phi(n)$ primitive nth roots of unity are precisely the set of roots of a single *irreducible* polynomial over $\mathbf{Q}$, with integral coefficients, called the *nth cyclotomic polynomial*, which thus has degree $\phi(n)$. But to show this is quite difficult (see Problem 5, Section 2.7), and so we adopt a simpler course here that does not depend on this result. In particular, we simplify the problem by looking at the divisors of n.

LEMMA 24a. *If the regular n-gon is constructible, then so too is the regular m-gon for any $m \geqslant 3$ such that $m \mid n$.*

Proof. If $n = mk$, just connect every kth vertex of the n-gon, beginning at any one. ∎

LEMMA 24b. *If the regular n-gon is constructible and if the odd prime p divides n, then p is of the form $2^{(2^k)} + 1$.*

Proof. By the hypothesis and Lemma 24a, the regular p-gon is constructible; and so a primitive pth root of unity is constructible. But the $p - 1$ primitive pth roots of unity are the roots of the polynomial $(x^p - 1)/(x - 1) = x^{p-1} + x^{p-2} + \cdots + 1$, which was shown in Problem 8 of Section 2.1 to be irreducible. By Theorem 23 then, $p - 1$ must be a power of 2; that is, $p = 2^m + 1$ for some positive integer m. However, the only time a number of this form can possibly be prime is when $m = 2^k$ for some k. To see this, note that for $m = 1$ we can take $k = 0$, and for $m > 1$, if m has an odd factor q, so that $m = qr$, then $2^m + 1 = [(2^r)^q + 1] = [2^r + 1][(2^r)^{q-1} - (2^r)^{q-2} + (2^r)^{q-3} - \cdots - 2^r + 1]$, which is not prime. ∎

Primes of the form $2^{(2^k)}+1$ are called *Fermat primes*; the only ones known to date are 3, 5, 17, 257, and 65537. Thus, if the regular n-gon is constructible, all the odd prime factors of n must be Fermat primes. Even further, no odd prime can divide n more than once, for we have:

LEMMA 24c. *If the regular n-gon is constructible and if p is an odd prime, then p^2 does not divide n.*

Proof. From Problem 4 of Section 2.4. we know that $\phi(p^2)=p(p-1)$. If $p^2 \mid n$, the $p(p-1)$ primitive p^2 roots of unity would have to be constructible. But these each have degree $p(p-1)$ over $\mathbf{Q}$, as was shown in Problem 10 of Section 2.4. Since their degree has the odd factor p, they cannot be constructible. ∎

Putting the previous two lemmas together, we immediately obtain the following necessary condition on n for the constructibility of the regular n-gon:

THEOREM 24. *For the regular n-gon to be constructible, n must be of the form $2^k p_1 p_2 \cdots p_m$, where the p_i's are distinct Fermat primes. (m may equal 0, in which case n has no odd factors.)*

As will be shown in Section 2.7, the converse of this theorem is also true. That is, for any n of the given form, it is indeed possible to construct the regular n-gon. But the proof of this is rather intricate and depends on certain number-theoretic results which we shall develop in the next section.

For the time being, we might observe that Theorem 24 implies that it is not possible to construct regular polygons with, for example, 7, 11, or 90 sides. For another look at the trisection problem, observe that the trisection of 60°

would necessitate the constructibility of a regular 18-gon; and since $18 = 2 \cdot 3^2$, this is impossible.

PROBLEMS

1. Give a proof of the multiplication and division parts of Lemma 22a using the polar representation of a_1 and a_2.
2. Give a proof of Lemma 22b which does not use the polar representation of a.
3. Give an analytic proof of Lemma 24a.
4. Prove that the regular pentagon is constructible.
5. Is it possible to divide an angle of 60° into five equal parts?

Section 2.6. Congruences

In this section, we shall develop some basic notions about congruences, which are a very important tool in number theory and in algebra. All the numbers referred to will be assumed to be integers.

If n is a positive integer, we say that *a is congruent to b modulo n*, written $a \equiv b \pmod{n}$, if $n \mid (a-b)$. This is easily seen to be equivalent to a and b having the same remainder when divided by n. For example, $21 \equiv 6 \pmod{5}$ since $5 \mid (21-6)$; equivalently, each has remainder 1 when divided by 5. Similarly, $-17 \equiv 3 \pmod{10}$ since $10 \mid (-17-3)$; equivalently, each has remainder 3 when divided by 10. (Review the division algorithm for **Z**, if necessary, in Problem 1 of Section 2.1.)

A statement of the form $a \equiv b \pmod{n}$ is called a *congruence*, and we shall now proceed to show that congruences behave very much like equations. In particular, if $a \equiv b \pmod{n}$ and $c \equiv d \pmod{n}$, then both

$a + c \equiv b + d \pmod{n}$ and $ac \equiv bd \pmod{n}$. These follow immediately from the definition, noting for the second case that $ac - bd = ac - ad + ad - bd = a(c - d) + d(a - b)$. Thus we can add (and subtract) and multiply congruences. Since we generally can't divide in $\mathbf{Z}$, we do not discuss the general division of one congruence by another. However, we do have the following *cancellation property*. Suppose that a and n are relatively prime, which we can abbreviate by writing the g.c.d. $(a, n) = 1$. Then the congruence $ab \equiv ac \pmod{n}$ implies that $b \equiv c \pmod{n}$, for if $n \mid a(b - c)$ but n has no factors in common with a, then $n \mid (b - c)$. This is completely analogous to the division of an ordinary equation by a nonzero factor of both sides. Lastly, congruences have the three properties of an *equivalence relation*: (i) for every a, $a \equiv a \pmod{n}$; (ii) if $a \equiv b \pmod{n}$, then $b \equiv a \pmod{n}$; and (iii) if $a \equiv b \pmod{n}$ and $b \equiv c \pmod{n}$, then $a \equiv c \pmod{n}$. These also follow at once from the definition.

The main value of congruences is the simple representation they afford for certain kinds of properties and calculations involving integers. For example, to show that the product of two odd numbers a and b is odd, we could represent a as $(2n + 1)$ and b as $(2m + 1)$ and calculate $ab = (2n + 1)(2m + 1) = 4nm + 2n + 2m + 1 = 2(2nm + n + m) + 1$, from which it follows that ab is odd. The same calculation, written in the language of congruences, might go like this: $a \equiv 1 \pmod{2}$, $b \equiv 1 \pmod{2}$, and hence $ab \equiv 1 \pmod{2}$. Much of the development of mathematical thought is due to the invention of simple notation that can be effectively manipulated; congruences are an excellent example of this.

It is not the purpose here to make a detailed study of congruences, which are ordinarily treated in number theory, but rather to develop a particular tool, Theorem

27, which will enable us to return effectively to the problem of the construction of the regular n-gon. The auxiliary results developed here will also be useful later on, and the reader should especially note some very close similarities between the statements and proofs here and those of earlier sections.

Frequent use will be made of the Euler ϕ-function; recall that $\phi(n)$ represents the number of integers between 1 and n (inclusive) which are relatively prime to n. To begin, we have **Fermat's Theorem**:

THEOREM 25. *If a and n are relatively prime, then $a^{\phi(n)} \equiv 1 \pmod{n}$.*

Proof. Let $r_1, r_2, \ldots, r_{\phi(n)}$ be the (distinct) numbers between 1 and n which are relatively prime to n. If s is any integer at all which is relatively prime to n, then for some unique i, $s \equiv r_i \pmod{n}$, for by the division algorithm we can write $s = nq + r$, and r cannot have a factor in common with n or else s would too. Thus $r =$ some r_i, and so $s \equiv r_i \pmod{n}$. In particular, each number of the form ar_i, since it has no factors in common with n, must satisfy $ar_i \equiv r_j \pmod{n}$ for some j. Further, if $i \neq k$, then $ar_i \not\equiv ar_k \pmod{n}$, for the contrary would imply that $r_i \equiv r_k \pmod{n}$, which is impossible since r_i and r_k are distinct numbers between 1 and n. Thus,

$$\prod_{i=1}^{\phi(n)} ar_i \equiv \prod_{j=1}^{\phi(n)} r_j \pmod{n},$$

$$a^{\phi(n)} \prod_{i=1}^{\phi(n)} r_i \equiv \prod_{j=1}^{\phi(n)} r_j \pmod{n}.$$

Since $\prod_{i=1}^{\phi(n)} r_i$ and n have no common factors, we conclude that $a^{\phi(n)} \equiv 1 \pmod{n}$. ∎

For p a prime, we know that $\phi(p) = p - 1$ and so in this important case we have the following immediate corollary:

COROLLARY 25. *If p is prime and if a is not divisible by p, then $a^{p-1} \equiv 1 \pmod{p}$. For any a, $a^p \equiv a \pmod{p}$.*

In the context of the first sentence of this corollary, if further $a^m \not\equiv 1 \pmod{p}$ for each m such that $1 \leqslant m \leqslant p-2$, then we say that a is a *primitive root modulo p*. (Watch the developing analogy, especially in the proofs, with primitive roots of unity.) It is not at all obvious that there will necessarily even exist a primitive root modulo p. In our proof of this fact, we shall need the following theorem:

THEOREM 26. *If $f(x) = a_n x^n + a_{n-1}x^{n-1} + \cdots + a_0$ is a polynomial with integral coefficients and if p is a prime which does not divide a_n, then the congruence $f(x) \equiv 0 \pmod{p}$ has at most n incongruent integral solutions.*

Proof. By induction on n. For $n = 0$, there are no solutions, since then $f(x) = a_0 \not\equiv 0 \pmod{p}$. Assuming that the theorem holds for all polynomials of degree $n-1$, we proceed to show it for each polynomial f of degree n. If the congruence $f(x) \equiv 0 \pmod{p}$ has no roots, then of course we are done. In the other case, we shall make use of the factorization, for any x and r,

$$\begin{aligned} f(x) - f(r) &= \sum_{k=1}^{n} a_k(x^k - r^k) \\ &= (x-r)\sum_{k=1}^{n} a_k(x^{k-1} + x^{k-2}r + x^{k-3}r^2 \\ &\qquad + \cdots + r^{k-1}) \\ &= (x-r)\,g(x), \end{aligned}$$

where g is a polynomial of degree $n-1$ with leading coefficient a_n, which is not divisible by p. Now if r is an integer, $g(x)$ has integral coefficients. If in addition r is a root of the congruence $f(x) \equiv 0 \pmod{p}$, we may rewrite the problem as

$$0 \equiv f(x) \equiv (x-r)\,g(x) \pmod{p}.$$

If x is any root incongruent to r, the cancellation property implies that $0 \equiv g(x) \pmod{p}$, and so, by the inductive hypothesis, there are at most $n-1$ roots incongruent to r. Thus, there are at most n incongruent solutions to the original congruence. ∎

Now we are able to settle the matter at hand, namely, the existence of primitive roots modulo p, for p a prime.

THEOREM 27. *If p is a prime, then there exist exactly $\phi(p-1)$ incongruent primitive roots modulo p.*

Proof. Consider the $p-1$ incongruent numbers $1, 2, \ldots, p-1$. In view of Fermat's Theorem, for each such number a there exists some minimal exponent $h \geqslant 1$ such that $a^h \equiv 1 \pmod{p}$. But h must in fact be a divisor of $p-1$, for if we apply the division algorithm for $\mathbf{Z}$ to obtain $p-1 = hq + r$, we see that $a^{p-1} = (a^h)^q a^r$, and so $a^r \equiv 1 \pmod{p}$, since a^{p-1} and a^h are both $\equiv 1 \pmod{p}$. By the minimality of h, $r = 0$; thus $h \mid (p-1)$.

For a fixed value of h dividing $p-1$, how many a's have this property, namely, that h is the *least* positive exponent for which $a^h \equiv 1 \pmod{p}$? (Such a's are said to *belong to the exponent h* modulo p, or to have *order h*.) If there is one such number a, the candidates must be restricted to the incongruent numbers $a, a^2, \ldots, a^h$, for these are each solutions to $x^h \equiv 1 \pmod{p}$; and since there are h of them, Theorem 26 guarantees that there cannot be any others. Even among these, the only possible candidates must be of the form a^k for k and h relatively prime. For if k and h have a common factor $q \geqslant 2$, then $(a^k)^{h/q} \equiv (a^h)^{k/q} \equiv 1 \pmod{p}$, so that a^k would have lower order than h. (Each of these candidates actually turns out to have order h, as will follow shortly.)

Thus, for each h, there are at most $\phi(h)$ incongruent numbers with order h. Letting $\psi(h)$ be the precise number of incongruent numbers with order h, we now have

$$p-1 = \sum_{h \mid p-1} \psi(h) \leqslant \sum_{h \mid p-1} \phi(h) = p-1,$$

where the last equality was proved in Problem 9 of Section 2.4. Thus, for each h, it must be that $\psi(h) = \phi(h)$, and so in particular, $\psi(p-1) = \phi(p-1)$. ∎

For example, for $p = 11$, there should be $\phi(10) = 4$ primitive roots. The table in Figure 13 gives the successive powers of each integer from 1 to 10, calculated modulo

	Exponents									
Numbers	1	2	3	4	5	6	7	8	9	10
1	1									
2	2	4	8	5	10	9	7	3	6	1
3	3	9	5	4	1					
4	4	5	9	3	1					
5	5	3	4	9	1					
6	6	3	7	9	10	5	8	4	2	1
7	7	5	2	3	10	4	6	9	8	1
8	8	9	6	4	10	3	2	5	7	1
9	9	4	3	5	1					
10	10	1								

Figure 13

11, until the number 1 is reached. Note that the four numbers 2, 6, 7, and 8 are primitive roots. Note also that the order of each number divides $\phi(11)$, and that for each h dividing $\phi(11)$, the number of numbers with this order is $\phi(h)$. These are illustrations of the facts developed in the proof of Theorem 27. Of course, instead of 2, 6, 7, and 8, we could have chosen any numbers congruent to them modulo 11, such as 13, -5, 95, and -36.

PROBLEMS

1. Use congruences to reformulate and prove the theorem that every perfect square is of the form $4n$ or $4n+1$.
2. If $a \not\equiv 0 \pmod{n}$ and $b \not\equiv 0 \pmod{n}$, does it follow that $ab \not\equiv 0 \pmod{n}$?
3. If a and n are relatively prime, show that there exists an integer b such that $ab \equiv 1 \pmod{n}$.
4. With the obvious definition, does there always exist a primitive root modulo n, when n is not prime?
5. Find all the primitive roots modulo 7.
6. How many primitive roots modulo 17 are there? Find one.

Section 2.7. Constructibility of Regular Polygons II

In Section 2.5 it was shown that the only values of n for which it might be possible to construct the regular n-gon are numbers of the form $2^k p_1 p_2 \cdots p_m$, where the p_i's are distinct Fermat primes. In this section, it will be shown that for all such numbers n, it actually is possible to construct the regular n-gon. The proof of this fact is rather intricate and may be skipped without loss of continuity. It has been included here because the ingenious argument, due to Gauss, is not readily found except in the original text, even though it is rather elementary. In Section 3.7 we shall be able to construct a very short proof, based on much more elaborate theory, and it will be interesting to relate that proof to the one given here. The proof of the theorem in this section is followed by a detailed example, and the reader may find it useful to consult the example before, during, and after the study of the proof.

To begin, the following two lemmas enable the general problem to be dealt with in simpler parts.

LEMMA 28a. *If m and n are relatively prime, and if both the regular m-gon and the regular n-gon are constructible, then the regular mn-gon is constructible.*

Proof. By the hypothesis, angles of $2\pi/m$ and $2\pi/n$ are constructible. We would like to show that an angle of $2\pi/(mn)$ is constructible. By the Euclidean algorithm for $\mathbf{Z}$, there exist integers s and t such that

$$1 = sm + tn$$

$$\frac{1}{mn} = \frac{s}{n} + \frac{t}{m}$$

$$\frac{2\pi}{mn} = s\frac{2\pi}{n} + t\frac{2\pi}{m}.$$

Therefore the angle $2\pi/(mn)$ is constructible, since it can be obtained as the sum of (possibly negative) multiples of constructible angles. (See Problem 3 for an analytic proof based on the roots of unity.) ∎

LEMMA 28b. *For $k \geqslant 2$, the regular 2^k-gon is constructible. For $k \geqslant 1$, if the regular n-gon is constructible, then so too is the regular $2^k n$-gon.*

Proof. The first sentence is obvious, requiring only repeated bisections of 360°. The second sentence is almost as obvious, requiring only repeated bisections of $360°/n$. ∎

And now we are ready for the main result:

THEOREM 28. *For $n \geqslant 3$ and of the form $2^k p_1 p_2 \cdots p_m$, where the p_i's are distinct Fermat primes, the regular n-gon is constructible.* (*m may equal* 0, *in which case n has no odd factors.*)

Proof. In view of Lemmas 28a and 28b, it suffices to treat the case $n = p$, where p is a Fermat prime. Hence we may write $p - 1 = 2^{(2^N)}$. We want to show that a primitive pth root of unity is constructible. Denoting such a root by ω, we recall that the entire set of primitive pth roots of unity is given by $\omega, \omega^2, \omega^3, \ldots, \omega^{p-1}$. From the equation $x^{p-1} + x^{p-2} + \cdots + x + 1 = 0$, which is satisfied by ω (as well as by the other primitive pth roots of unity), we conclude that the sum of all the primitive pth roots of unity is -1.

In the course of our computations we shall frequently make use of the fact that $\omega^i = \omega^j$ if and only if $i \equiv j$ (mod p), which follows at once from the definition of ω.

Let g be a primitive root modulo p; we know that such primitive roots exist by Theorem 27. Since the numbers $g, g^2, g^3, \ldots, g^{p-1}$ are all incongruent, and since none is congruent to 0, the primitive pth roots of unity may also be represented by $\omega^g, \omega^{g^2}, \omega^{g^3}, \ldots, \omega^{g^{p-1}}$. Because the exponents here and in what follows are frequently themselves exponentiated quantities, we shall adopt the abbreviation $[k]$ for ω^k. Thus the set S of primitive pth roots of unity may be listed as $[g], [g^2], \ldots, [g^{p-1}]$. In referring to the elements of the set S as well as to the elements of other sets to be defined below, it will be convenient to make reference to the order in which these elements are listed. It is important to note that each element in this list is the gth power of the one before it, and that the first element is the gth power of the last. Consequently, we could list the primitive pth roots of unity by starting with *any* element in the list and successively raising it to the gth power until we had generated the whole list. For example, starting with $[g^j]$ and doing this, we get $[g^j], [g^{j+1}], [g^{j+2}], \ldots, [g^{p-1}], [g], \ldots, [g^{j-1}]$, where some of the middle terms have been included to clarify what is going on. If we were to continue raising these elements to the gth power, the list would just start repeating itself.

Let us denote by S_1 the set $[g], [g^3], [g^5], \ldots, [g^{p-2}]$, where we have taken the first and every other element of S. Similarly, denote by S_2 the set $[g^2], [g^4], [g^6], \ldots, [g^{p-1}]$. Each of these sets may be generated by taking successive g^2 powers of any one of its elements. Indeed they could even be defined in this way. We could define S_1 as the set generated by all the g^2 powers of the first element of S and S_2 as the set generated by all the g^2 powers of the second element of S. (As with S, after going through sufficiently many g^2 powers, the lists would just start to repeat.) It may occasionally be helpful to think of S_1 and S_2 as having been formed by taking 'every other' element of S.

Now, for $i = 1, 2$, we subdivide S_i into sets $S_{i,1}$ and $S_{i,2}$ by an analogous process. In particular $S_{i,1}$ would be generated by successive g^4 powers of the first element of S_i, which has the effect again of taking every other element; likewise $S_{i,2}$ would be generated by successive g^4 powers of the second element of S_i.

We should like to continue repeating this process of dividing sets in half until each set contains one single element. To give a precise inductive definition of the process, some more terminology is needed. An *m-set* is a set resulting from m divisions; it can be denoted by m subscripts, although in dealing with a particular m-set we may agree to suppress the writing of the subscripts. Thus S is a 0-set, S_1 and S_2 are 1-sets, and $S_{1,1}$, $S_{1,2}$, $S_{2,1}$, and $S_{2,2}$ are 2-sets. Each m-set may be generated by successive g^{2^m} powers of any of its elements; and each such set contains $(p-1)/2^m$ elements. Each m-set, $S_{i_1, i_2, \ldots, i_m}$, gives rise to two $(m+1)$-sets, $S_{i_1, i_2, \ldots, i_m, 1}$ and $S_{i_1, i_2, \ldots, i_m, 2}$ in the natural way. That is, $S_{i_1, i_2, \ldots, i_m, 1}$ is the set generated by successive $g^{(2^{m+1})}$ powers of the first element of $S_{i_1, i_2, \ldots, i_m}$. Similarly, $S_{i_1, i_2, \ldots, i_m, 2}$ is the set generated by successive $g^{(2^{m+1})}$ powers of the second element of $S_{i_1, i_2, \ldots, i_m}$. The net effect, again, is to pick out every other element. The two $(m+1)$-sets resulting from a given m-set are said to be *complementary*.

It is important to observe that the fact that $p-1$ is a power of 2 is critical to this procedure. If $p-1$ were not such, then the repeated division into two equal subsets would not be possible, nor would the representation of sets as successive g^{2^m} powers of arbitrary elements. In the present case, after $m = 2^N$ divisions, each m-set contains a single element, namely, some primitive pth root of unity.

We denote by the *period* of a set the sum of its elements. It is an *m-period* if it is the period of an m-set. Two periods are *complementary* if they are the periods of complementary sets. Thus, each m-period is the sum of some $(p-1)/2^m$ primitive pth roots of unity; and, in particular, when $m = 2^N$, each such period is precisely one of the primitive pth roots of unity. Thus, *the theorem will be proved if we can show that for* $0 \leqslant m \leqslant 2^N$, *each m-period is constructible.* We shall show this by induction on m, the essence of the situation being that each $(m+1)$-period is the root of a quadratic equation whose coefficients are linear combinations of m-periods. (In fact, complementary periods are the two roots of the same quadratic equation.)

To prepare for the argument, recall that if we know the sum and product of two numbers: say $x_1 + x_2 = A$ and $x_1 x_2 = B$, then x_1 and x_2 are the roots of the quadratic $x^2 - Ax + B$. (Cf. Section 1.6.) Consequently, if A and B are constructible, then so too are x_1 and x_2, since they may be obtained from A and B via the quadratic formula, in which the operations correspond to admissible constructions.

We now enter into the main part of the proof that for each m, $0 \leqslant m \leqslant 2^N$, all the m-periods are constructible. For $m = 0$, the only 0-period is the sum of all the primitive pth roots, and it has already been noted that this sum is simply -1, which is, of course, constructible. Let us assume that all periods through the $(m-1)$-periods are constructible. Let η_1 be an arbitrary m-period and η_2 its complementary period. By the previous paragraph, it suffices to show that $\eta_1 + \eta_2$ and $\eta_1\eta_2$ are constructible. Since $\eta_1 + \eta_2$ is precisely one of the $(m-1)$-periods, by the inductive hypothesis it is constructible. The constructibility of $\eta_1\eta_2$ will follow rather easily from the following claim.

Claim. For $1 \leqslant m \leqslant 2^N - 1$, $\eta_1\eta_2$ may be expressed as a linear combination, with non-negative integral coefficients, of all the m-periods. In this linear combination, complementary periods have equal coefficients. For $m = 2^N$, $\eta_1\eta_2 = 1$.

Proof. η_1 and η_2 are the sums of the elements in two complementary m-sets S' and S''. These two m-sets resulted from the division of some one $(m-1)$-set, which is generated by successive $g^{(2^{m-1})}$ powers of some particular element ω^k. Letting $h=g^{(2^{m-1})}$, this $(m-1)$-set may be written $[k], [kh], [kh^2], \ldots, [kh^{f-2}]$, where for convenience we have defined $f=1+(p-1)/2^{m-1}$. Subsequent powers of h just repeat the list, for $h^{f-1}=(g^{(2^{m-1})})^{(p-1)/2^{m-1}}=g^{p-1}\equiv 1 \pmod p$.

Thus we have

$$\eta_1=[k]+[kh^2]+\cdots+[kh^{f-3}],$$
$$\eta_2=[kh]+[kh^3]+\cdots+[kh^{f-2}].$$

For $m=2^N$, $h=g^{(p-1)/2}\equiv -1 \pmod p$, so that $\eta_1\eta_2=\omega^k\omega^{-k}=1$. For $m<2^N$, we write out the product $\eta_1\eta_2$, grouping the terms appropriately (and remembering that $[kh^i][kh^j]=[kh^i+kh^j]$ since $[t]$ denotes ω^t):

$$\begin{aligned}\eta_1\eta_2 = {} & [k+kh] && +[kh^2+kh^3] && +\cdots+[kh^{f-3}+kh^{f-2}] \\ & +[k+kh^3] && +[kh^2+kh^5] && +\cdots+[kh^{f-3}+kh] \\ & +[k+kh^5] && +[kh^2+kh^7] && +\cdots+[kh^{f-3}+kh^3] \\ & \vdots \\ & +[k+kh^{f-4}] && +[kh^2+kh^{f-2}] && +\cdots+[kh^{f-3}+kh^{f-6}] \\ & +[k+kh^{f-2}] && +[kh^2+kh] && +\cdots+[kh^{f-3}+kh^{f-4}].\end{aligned}$$

Careful scrutiny of this sum will lead to the conclusions in the claim. First, each row of the sum is generated by the h^2 powers of any of its elements. Second, none of the exponents is congruent to 0 modulo p. To see the latter, it suffices to treat the first column. Suppose $k + kh^{2j+1} \equiv 0 \pmod p$, where $1 \leqslant 2j + 1 \leqslant f - 2$. Then we would have

$$-1 \equiv h^{2j+1} \pmod p$$

$$1 \equiv h^{4j+2} \pmod p$$

$$1 \equiv g^{(2^{m-1})(4j+2)} \pmod p,$$

and so, since g is a primitive root modulo p, $(p-1) \mid (2^{m-1})(4j+2)$. But

$$2 \leqslant 4j + 2 \leqslant 2f - 4 = \frac{(p-1)}{2^{m-2}} - 2,$$

and so

$$\begin{aligned}(2^{m-1})(4j+2) &\leqslant 2(p-1) - 4 \\ &< 2(p-1).\end{aligned}$$

Consequently, we would have to have

$$\begin{aligned}p - 1 &= (2^{m-1})(4j+2) \\ &= 2^m(2j+1).\end{aligned}$$

Since $p - 1$ has no odd factors, the only value of j for which this might hold is $j = 0$; but then we would have $p - 1 = 2^m \leqslant 2^{(2^N - 1)} = (p-1)/2$, which is impossible. Consequently *each row is some m-period*, although not necessarily $\eta_1\eta_2$. (Remember, there are 2^m m-sets, each with a corresponding period.) Furthermore, the first and the last row are complementary m-periods, for the hth power of any element of the first row is an element of the last row. For example, taking the hth power of the first element of the first row results in:

$$[k+kh]^h = [kh+kh^2] \qquad \text{(law of exponents)}$$
$$= [kh^2+kh]$$

which is the second element of the last row. Since the other terms in these rows are generated by successive h^2 powers of these, it suffices simply to find two terms related in this way, as has been done, Similarly, it is easy to see that the hth power of the first entry in the second row is the third entry in the next-to-last row, and so these rows are complementary. By analogous argument, which we do not formalize, it is seen that the ith row from the top is complementary to the ith row from the bottom. Since the number of rows is even, being $(f-1)/2$, the rows can all be paired up in this way, and so each m-period occurs as often as its complement. This completes the proof of the claim.

By the claim, $\eta_1\eta_2$ equals a sum of terms of the form $R(\eta+\tilde{\eta})$, where η and $\tilde{\eta}$ are complementary m-periods and R is a non-negative integer. But since η and $\tilde{\eta}$ are complementary, $\eta+\tilde{\eta}$ is just some $(m-1)$-period, and hence constructible according to the inductive hypothesis. Being a sum of constructible numbers, $\eta_1\eta_2$ must then itself be constructible.

Having shown that $\eta_1+\eta_2$ and $\eta_1\eta_2$ are constructible, we are able to conclude (as noted at the outset) that η_1 and η_2 are each constructible, which completes the proof. ∎

As an example, let us reconstruct the proof of Theorem 28 for the case $p = 17$. If ω is any primitive 17th root of unity, then all the primitive 17th roots of unity are given by $\omega, \omega^2, \ldots, \omega^{16}$. But if g is any primitive root modulo 17, we can also write this list as $\omega^g, \omega^{g^2}, \ldots, \omega^{g^{16}}$, which will simply be a reordering of the previous list. Taking $g = 3$, which is a primitive root modulo 17, the resulting list is $\omega^3, \omega^9, \omega^{27}, \omega^{81}, \ldots, \omega$, which is made simpler by reducing each of their exponents modulo 17. In this way we obtain the following complete list: ω^3, ω^9, ω^{10}, ω^{13}, ω^5, ω^{15}, ω^{11}, ω^{16}, ω^{14}, ω^8, ω^7, ω^4, ω^{12}, ω^2, ω^6, ω^1. We call this list S.

In the course of the computations to follow, it is possible to adopt two courses. First, we could show that all the periods can be obtained as the roots of successive quadratic equations, from which the constructibility of the periods, and hence of the primitive 17th roots, follows. This is all that was necessary and all that was done in the proof of Theorem 28. However, this course does not yield specific directions on how to construct the polygon because it does not tell which root of each quadratic corresponds to which period. For example, we may be able to find a quadratic satisfied by two periods η_1 and η_2, but further analysis is necessary to determine which root corresponds to η_1 and which corresponds to η_2.

For the purposes of this example, let us treat also this second problem, so that at the end we shall actually have a definite algorithm for constructing the regular 17-gon. Our analysis of this aspect will be geometrical and somewhat intuitive, but it should be clear how the steps could easily be translated into verifiable inequalities involving cosines. For definiteness, we take $\omega = e^{2\pi i/17}$. A sketch of the location of the powers of ω is given in Figure 14, where the point ω^k is labeled by the exponent k. This

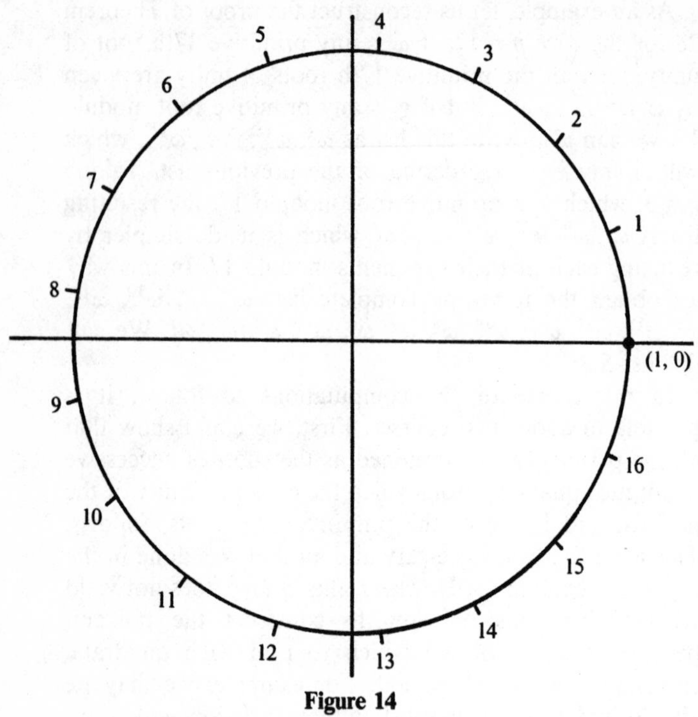

Figure 14

sketch, to which we shall make frequent reference, does not represent a *construction*; but we do know the numerical values of the relevant central angles as multiples of $2\pi/17$.

Now we proceed along the lines of the proof of Theorem 28. The set S has period $\eta = -1$, since the sum of the primitive pth roots of unity is -1. S is now divided into two sets S_1 and S_2, with corresponding periods:

$$\eta_1 = \omega^3 + \omega^{10} + \omega^5 + \omega^{11} + \omega^{14} + \omega^7 + \omega^{12} + \omega^6,$$

$$\eta_2 = \omega^9 + \omega^{13} + \omega^{15} + \omega^{16} + \omega^8 + \omega^4 + \omega^2 + \omega^1.$$

Clearly $\eta_1 + \eta_2 = \eta = -1$. By calculation we find $\eta_1\eta_2 = -4$, which the reader might care to verify. Thus η_1 and η_2 are the roots of the quadratic $x^2 + x - 4$. Now we need to see which root is η_1 and which is η_2.

Since the summands in each period can be paired off into pairs of complex conjugates (ω^3 and ω^{14}, ω^{10} and ω^7, etc.), the periods may be rewritten as

$$\eta_1 = 2\,\mathrm{Re}(\omega^3 + \omega^5 + \omega^6 + \omega^7),$$

$$\eta_2 = 2\,\mathrm{Re}(\omega^1 + \omega^2 + \omega^4 + \omega^8),$$

where Re denotes the real part of the given number. (Thus the periods are real.) From the geometrically evident inequalities,

$$\mathrm{Re}(\omega^2) - \mathrm{Re}(\omega^3) > \mathrm{Re}(\omega^7) - \mathrm{Re}(\omega^8),$$

$$\mathrm{Re}(\omega^1) > \mathrm{Re}(\omega^5),$$

$$\mathrm{Re}(\omega^4) > \mathrm{Re}(\omega^6),$$

which could also be established analytically, we see that

$\eta_2 > \eta_1$. Thus

$$\eta_1 = \frac{-1-\sqrt{17}}{2},$$

$$\eta_2 = \frac{-1+\sqrt{17}}{2}.$$

Proceeding to the next stage, we divide S_1 into $S_{1,1}$ and $S_{1,2}$, and S_2 into $S_{2,1}$ and $S_{2,2}$; the corresponding periods are

$$\eta_{1,1} = \omega^3 + \omega^5 + \omega^{14} + \omega^{12}$$

$$\eta_{1,2} = \omega^{10} + \omega^{11} + \omega^7 + \omega^6$$

$$\eta_{2,1} = \omega^9 + \omega^{15} + \omega^8 + \omega^2$$

$$\eta_{2,2} = \omega^{13} + \omega^{16} + \omega^4 + \omega^1.$$

We see that $\eta_{1,1} + \eta_{1,2} = \eta_1$, which is known. By a calculation, $\eta_{1,1}\eta_{1,2} = -1$. Thus, these two periods are the roots of $x^2 - \eta_1 x - 1$. By an analysis similar to that in the preceding paragraph, we determine which root is which. In particular,

$$\eta_{1,1} = 2\ \mathrm{Re}(\omega^3 + \omega^5)$$

$$\eta_{1,2} = 2\ \mathrm{Re}(\omega^6 + \omega^7),$$

from which we see, using the diagram, that $\eta_{1,1} > \eta_{1,2}$. Therefore,

$$\eta_{1,1} = \frac{\eta_1 + \sqrt{\eta_1^2 + 4}}{2},$$

$$\eta_{1,2} = \frac{\eta_1 - \sqrt{\eta_1^2 + 4}}{2}.$$

Similarly, one finds that

$$\eta_{2,1} = \frac{\eta_2 - \sqrt{\eta_2^2 + 4}}{2},$$

$$\eta_{2,2} = \frac{\eta_2 + \sqrt{\eta_2^2 + 4}}{2}.$$

It is perhaps worth noting that the products of the complementary 2-periods, as obtained in this step, do not contain explicitly the 1-periods, as appears in the proof of the theorem. It simply happens to be the case here that the 1-periods have the same coefficients themselves, and so the product is given in terms of the 0-period, $\eta = -1$.

The subsequent division of $S_{1,1}$, $S_{1,2}$, $S_{2,1}$, and $S_{2,2}$ results in the eight 3-periods:

$$\begin{aligned}
\eta_{1,1,1} &= \omega^3 + \omega^{14} & \eta_{2,1,1} &= \omega^9 + \omega^8 \\
\eta_{1,1,2} &= \omega^5 + \omega^{12} & \eta_{2,1,2} &= \omega^{15} + \omega^2 \\
\eta_{1,2,1} &= \omega^{10} + \omega^7 & \eta_{2,2,1} &= \omega^{13} + \omega^4 \\
\eta_{1,2,2} &= \omega^{11} + \omega^6 & \eta_{2,2,2} &= \omega^{16} + \omega^1.
\end{aligned}$$

Now we don't really have to go to the trouble of finding all these; it turns out that it will be enough to find the last two, $\eta_{2,2,1}$ and $\eta_{2,2,2}$. Their sum is $\eta_{2,2}$, which is already known; and their product is $\eta_{1,1}$, which is also already known. Thus they are the roots of $x^2 - \eta_{2,2}x + \eta_{1,1}$. From the observation that $\eta_{2,2,2} > \eta_{2,2,1}$, we have

$$\eta_{2,2,1} = \frac{\eta_{2,2} - \sqrt{\eta_{2,2}^2 - 4\eta_{1,1}}}{2},$$

$$\eta_{2,2,2} = \frac{\eta_{2,2} + \sqrt{\eta_{2,2}^2 - 4\eta_{1,1}}}{2}.$$

The last pair of 4-periods are simply ω^{16} and ω^{1}. Their sum is $\eta_{2,2,2}$ and their product is 1. We find then that

$$\omega = \frac{\eta_{2,2,2} + \sqrt{\eta_{2,2,2}^2 - 4}}{2}.$$

Having now ω, we have of course essentially all its powers as well.

In summary, the construction of ω requires the construction of the following sequence of numbers:

$$\eta_1 = \frac{-1 - \sqrt{17}}{2}$$

$$\eta_2 = \frac{-1 + \sqrt{17}}{2}$$

$$\eta_{1,1} = \frac{\eta_1 + \sqrt{\eta_1^2 + 4}}{2}$$

$$\eta_{2,2} = \frac{\eta_2 + \sqrt{\eta_2^2 + 4}}{2}$$

$$\eta_{2,2,2} = \frac{\eta_{2,2} + \sqrt{\eta_{2,2}^2 - 4\eta_{1,1}}}{2}$$

$$\omega = \frac{\eta_{2,2,2} + \sqrt{\eta_{2,2,2}^2 - 4}}{2}.$$

Thus we have given an actual algorithm for the construction of the regular 17-gon. The reader may wish to go ahead and carry it out, but the author does not have the intestinal fortitude!

PROBLEMS

1. Carry through the proof of Theorem 28 for the case $p = 5$, and deduce from it an algorithm for the construction of the regular pentagon.

2. Generalize the method of this section to show that the primitive seventh roots of unity may be obtained by the successive solution of polynomials of degrees 2 and 3. (You are not asked to solve these lower degree polynomials.)

3. If a is a primitive mth root of unity and if b is a primitive nth root of unity, and if $(m, n) = 1$, show that ab is a primitive mnth root of unity. Use this fact to give an alternate proof of Lemma 28a.

*4. Find a necessary and sufficient condition on a rational number α such that an angle of α degrees is constructible.

*5. Show that all the $\phi(n)$ primitive nth roots of unity are the roots of a single irreducible polynomial over $\mathbf{Q}$ which is monic and has integral coefficients. Use this result to give an alternate proof of Theorem 24.

REFERENCES AND NOTES

Further discussion about basic techniques and concepts used in this chapter may be found as follows: divisibility properties of polynomials and their relation with those of the integers, [1, 2] and textbooks on abstract algebra, such as [3, 4]; vector spaces and linear algebra, [5, 6]; number theoretic techniques such as congruences and the Euler ϕ-function, [7, 8]. The sufficient condition for constructibility of regular n-gons is due to Gauss [9], who made a significant step with the construction of the regular 17-gon (see [10]). Gauss claimed in [9] to have proved that the condition was also necessary, but he did not present it. Credit now generally is given to Wantzel [11] for this part, which may then also be considered to be essentially the first proof that an arbitrary angle cannot be trisected. For further discussion see [12, 13], the latter noting a valiant ten-year project of Hermes to construct a regular 65537-gon (?). There is a long history of proofs of the irreducibility of the nth cyclotomic polynomial. An excellent survey is contained in Section 13 of [14]. The proof in our solution to Problem 5, Section 2.7, comes from [3]; it is based on a famous eight-line proof by Landau [15]. Concise versions phrased in terms of polynomials over finite fields are in [16, 17].

1. N. H. McCoy, *Rings and Ideals*, Carus Mathematical Monographs, Number Eight, Mathematical Association of America, 1948.

2. Abraham Robinson, *Numbers and Ideals*, Holden-Day, San Francisco, 1965.

3. Allan Clark, *Elements of Abstract Algebra*, Wadsworth, Belmont, Calif., 1971.

4. Abraham P. Hillman and Gerald L. Alexanderson, *A First Undergraduate Course in Abstract Algebra*, Wadsworth, Belmont, Calif., 1973.

5. Daniel T. Finkbeiner, II, *Introduction to Matrices and Linear Transformations*, 2nd ed., Freeman, San Francisco, 1966.

6. Evar D. Nering, *Linear Algebra and Matrix Theory*, 2nd ed., Wiley, New York, 1970.

7. Ivan Niven and Herbert S. Zuckerman, *An Introduction to the Theory of Numbers*, 3rd ed., Wiley, New York, 1972.

8. Edmund Landau, *Elementary Number Theory*, tr. by J. Goodman, Chelsea, New York, 1958 (orig. ed., 1927).

9. Carl Friedrich Gauss, *Disquisitiones Arithmeticae*, tr. by A. Clark, Yale University Press, New Haven, 1966 (orig. ed., 1801).

10. R. C. Archibald, Gauss and the regular polygon of seventeen sides, *Amer. Math. Monthly*, 27 (1920) 323–326.

11. P. L. Wantzel, Recherches sur les moyens de reconnaître si un problème de géométrie peut se résoudre avec la règle et le compas, *J. Math. Pures Appl.*, 2 (1837) 366–372.

12. James Pierpont, On an undemonstrated theorem of the *Disquisitiones Arithmeticae, Bull. Amer. Math. Soc.*, 2 (1895) 77–83.

13. H. S. M. Coxeter, *Introduction to Geometry*, 2nd ed., Wiley, New York, 1969.

14. L. E. Dickson, *et al.*, *Algebraic Numbers*, Chelsea, New York; orig. pub. as *Report of the Committee on Algebraic Numbers*, National Research Council, 1923–1928.

15. E. Landau, Über die Irreduzibilität der Kreisteilungsgleichung, *Math. Z.*, 29 (1929) 462.

16. B. L. van der Waerden, *Modern Algebra*, vol. 1, Frederick Ungar, New York, 1949 (orig. ed. 1931).

17. Nathan Jacobson, *Basic Algebra I*, Freeman, San Francisco, 1974.

CHAPTER 3

SOLUTION BY RADICALS

Section 3.1. Statement of the Problem

As every student of mathematics knows, the solutions to the quadratic equation $ax^2 + bx + c = 0$ are given by the formula $x = (-b \pm \sqrt{b^2 - 4ac})/2a$. In this chapter we essentially ask: **To what extent can similar formulas be found for polynomials of higher degree?**

To begin, the term "similar formula" needs to be given precise meaning. For our purposes, we shall consider a formula to be similar if it involves only a sequence of rational operations and the extraction of roots (of any integral order), beginning with the coefficients. Since, beginning with any nonzero coefficient, any rational number may be computed by a sequence of rational operations, certainly any rational constants may be used at any point in the procedure, as are 2 and 4 in the quadratic formula as given above.

Even the word "formula" requires clarification, however. We shall use it to mean some definite procedure which may be applied to a whole class of problems. For the quadratic case, the usual formula covers all quadratic polynomials. It turns out that there are similar formulas, albeit more complicated, for cubic and quartic polynomials. (See Problems 1 and 4.) But for polynomials of higher degree the situation is entirely different, for it will be shown that for each $n \geqslant 5$, there exist polynomials of degree n which have roots that cannot be calculated by *any* sequence of rational operations and the extraction of roots, beginning with the coefficients!

For the sake of definiteness, let us take a look at one of these polynomials, namely, $f(x) = 2x^5 - 5x^4 + 5$. The graph of this polynomial for real values of x is sketched in Figure 15. There is a local minimum at $(2, -11)$ and a local maximum at $(0, 5)$. It is easy to see that there are three real roots and hence two complex, non-real, ones. *These roots are definite numbers*, but, as will be shown later, not a single one of them can be computed by any sequence of rational operations or the extraction of roots, beginning with the coefficients!

Thus, not only would it be hopeless to ask for such a formula that would work for all polynomials of degree five, say, but it would even be hopeless to ask for a formula that would work for the single polynomial $2x^5 - 5x^4 + 5$. Let us say that a polynomial is *solvable by radicals* if for each of its roots r there exists a sequence of rational operations and the extraction of roots through which r can be obtained from the coefficients. In light of our original question, it is now natural to ask: **When is a polynomial solvable by radicals?**

This is a very deep question, which will finally be answered in Section 3.5. To get started, it is convenient to rephrase the definition of solvability by radicals in terms of field extensions. If f is a polynomial over a field $\mathbf{F}$, the *splitting field of f, over* $\mathbf{F}$, is the smallest field containing $\mathbf{F}$ and all the roots of f. A field $\mathbf{E}$ is a *radical extension* of a field $\mathbf{F}$ if $\mathbf{E} = \mathbf{F}(k)$ for some $k \notin \mathbf{F}$ such that $k^n \in \mathbf{F}$ for some positive integer n. That is, $\mathbf{E}$ is the simple algebraic extension of $\mathbf{F}$ by some nth root of an element of $\mathbf{F}$. We

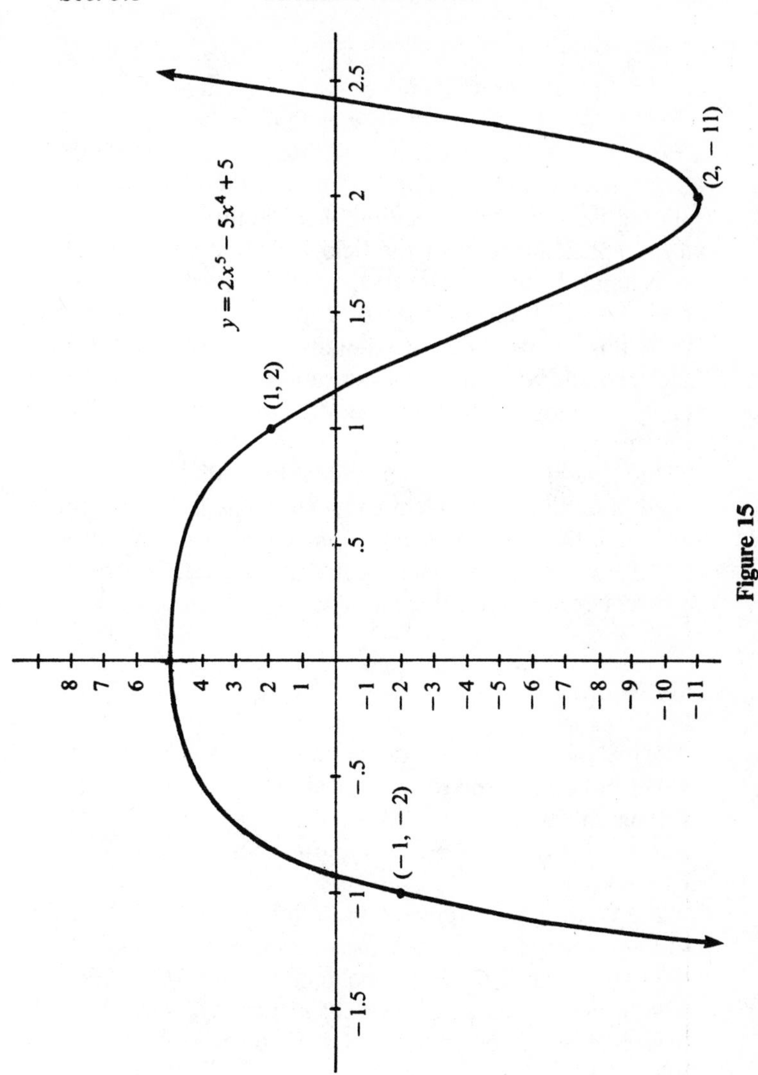

Figure 15

shall say that a polynomial $f \in \mathbf{F}[x]$ is *solvable by radicals over* $\mathbf{F}$ if and only if there is a sequence of fields $\mathbf{F} = \mathbf{F}_0 \subset \mathbf{F}_1 \subset \mathbf{F}_2 \subset \cdots \subset \mathbf{F}_N$ such that for each j, $\mathbf{F}_{j+1}$ is a radical extension of $\mathbf{F}_j$, and such that the splitting field $\mathbf{E}$, of f, over $\mathbf{F}$, satisfies $\mathbf{E} \subset \mathbf{F}_N$. In both the definition of splitting field and the definition of solvability by radicals, when explicit mention of the field $\mathbf{F}$ is omitted, we shall understand $\mathbf{F}$ to be the smallest field containing the coefficients of f; this field is called the *coefficient field of f*. With this convention, it should be clear that this field-theoretic definition of solvability by radicals is simply a reformulation of the definition given in the previous paragraph.

Thus, the problem of solvability by radicals can be stated in terms of a problem about field extensions. When is the splitting field of a polynomial contained in some field $\mathbf{F}_N$ which can be obtained from the coefficient field by a sequence of radical extensions?

PROBLEMS

1. Show that the general cubic equation $ax^3 + bx^2 + cx + d = 0$ can be solved by radicals. (Hint: Without loss of generality, take $a = 1$. Make the successive substitutions $x = y + L$ and $y = z + K/z$ for appropriate choices of L and K.)

2. What is the degree of the splitting field of $x^3 - 2$ as an extension of $\mathbf{Q}$? Find a basis for this extension.

3. Find an irreducible polynomial over $\mathbf{Q}$ whose splitting field has degree less than $n!$ over $\mathbf{Q}$, where n is the degree of the polynomial. (Cf. Problem 8 of Section 2.2.)

*4. Show that the general quartic equation $ax^4 + bx^3 + cx^2 + dx + e = 0$ may be solved by radicals. (Hint:

Without loss of generality, take $a = 1$. Let $y = x + (b/4)$, and then try to write the resulting equation as a difference of squares, making use of some extra parameter.)

5. Show that there exists a number which has degree 4 over **Q** but which is not constructible. (Hint: Consider the roots of $f(x) = 4x^4 - 4x + 3$.)

Section 3.2. Automorphisms and Groups

As noted at the conclusion of the previous section, we are faced with a problem about fields. **When is the splitting field of a polynomial contained in some field which can be obtained from the coefficient field by a finite sequence of radical extensions?** Fields are fairly complicated mathematical objects, containing in all cases under consideration here an infinite number of elements, which can be combined by means of two arithmetic operations (addition and multiplication) and their corresponding 'inverse' operations (subtraction and division). The power of the theory evolving from the work of Galois derives from the way it enables us to rephrase our problem about fields into a problem about much simpler mathematical objects, called groups, which have only a single arithmetic operation and, in the present case, only a finite number of elements.

The well-known mathematician Emil Artin (1898–1962) once wrote: "In modern mathematics, the investigation of the symmetries of a given mathematical structure has always yielded the most powerful results." We shall see this theme borne out in the developments of this chapter. To begin, let us formulate an interpretation of the word "symmetries". Most reasonable people would probably agree that the square is more 'symmetrical' than the general rectangle, both of which are sketched in Figure 16.

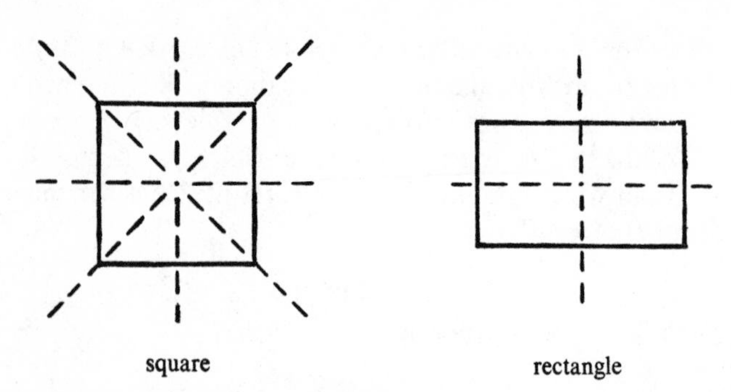

Figure 16

If asked to support this point of view, one might respond that there are *eight* distinct rigid motions of the square that will map it onto itself (reflections in each of the four dotted lines as well as rotations of 0°, 90°, 180°, and 270° about its center in, say, the counterclockwise direction), whereas for the rectangle there are only *four* (reflections in each of the two dotted lines as well as rotations of 0° and 180° about its center). Thus it is quite natural to measure the symmetry of an object by looking at the different ways in which it can be mapped onto itself so as to preserve its structure, and this is precisely what we shall do with field extensions.

If $\mathbf{F}$ is a field and if ϕ is a function mapping $\mathbf{F}$ to $\mathbf{F}$, we say that ϕ is an *automorphism* of $\mathbf{F}$ if it is one-to-one, onto, and it 'preserves the field operations'. By this latter phrase we mean that for any numbers $a, b \in \mathbf{F}$, $\phi(a+b) = \phi(a)+\phi(b)$, $\phi(a-b)=\phi(a)-\phi(b)$, $\phi(ab)=\phi(a)\phi(b)$, and $\phi(a/b)=\phi(a)/\phi(b)$. It is rather simple to see (Problem 1) that any automorphism must leave each rational number unchanged. Furthermore, if a function ϕ from $\mathbf{F}$ to $\mathbf{F}$ is one-to-one and onto and preserves addition and multiplication, then in fact it must preserve also subtraction and division, and thus be an automorphism. We shall make frequent use of these facts.

Clearly the identity mapping, which maps each element of **F** to itself, is an automorphism. As another example, the mapping of **C** given by complex conjugation, $\phi(a+bi)=a-bi$, a and b real, is an automorphism of **C**. To verify this, since ϕ is obviously one-to-one and onto, it suffices to show that ϕ preserves addition and multiplication. We have

$$\begin{aligned}\phi[(a+bi)+(c+di)] &= \phi[(a+c)+(b+d)i]\\ &= (a+c)-(b+d)i\\ &= (a-bi)+(c-di)\\ &= \phi(a+bi)+\phi(c+di),\\ \phi[(a+bi)(c+di)] &= \phi[(ac-bd)+(bc+ad)i]\\ &= (ac-bd)-(bc+ad)i\\ &= (a-bi)(c-di)\\ &= \phi(a+bi)\phi(c+di),\end{aligned}$$

which completes the verification. More examples will appear as we proceed.

Since automorphisms are functions from **F** to **F**, it makes sense to form the composition $\phi \circ \psi$ of two such automorphisms. For brevity we usually write this new function as $\phi\psi$; we have $\phi\psi(a)=\phi(\psi(a))$ for every $a \in \mathbf{F}$. The function $\phi\psi$ is frequently called the *product* of ϕ and ψ. For functions in general, the order in which they are composed makes quite a difference, and this is the case for automorphisms as well. In the exceptional circumstance when $\phi\psi=\psi\phi$, we say that ϕ and ψ *commute*. At the same time, the composition of functions is always *associative*, so we always have $(\phi\psi)\tau = \phi(\psi\tau)$, where ϕ, ψ and τ are automorphisms, and thus we can write the product $\phi\psi\tau$ unambiguously. Since an automorphism ϕ is one-to-one and onto, it has a unique inverse function, which we abbreviate ϕ^{-1}, and which satisfies $\phi\phi^{-1} = \phi^{-1}\phi =$ identity mapping on $\mathbf{F}$.

LEMMA 29a. *If ϕ and ψ are automorphisms of a field $\mathbf{F}$, then so too are ϕ^{-1} and $\phi\psi$.*

Proof. ϕ^{-1} is obviously one-to-one and onto. We want to show that it preserves addition and multiplication. We have

$$\begin{aligned}\phi\left[\phi^{-1}(a+b)\right] = a+b &= \phi\left[\phi^{-1}(a)\right] + \phi\left[\phi^{-1}(b)\right] \\ &= \phi\left[\phi^{-1}(a) + \phi^{-1}(b)\right]\end{aligned}$$

so by the fact that ϕ is one-to-one,

$$\phi^{-1}(a+b) = \phi^{-1}(a) + \phi^{-1}(b).$$

Similarly,

$$\begin{aligned}\phi\left[\phi^{-1}(ab)\right] = ab &= \phi\left[\phi^{-1}(a)\right] \cdot \phi\left[\phi^{-1}(b)\right] \\ &= \phi\left[\phi^{-1}(a) \cdot \phi^{-1}(b)\right]\end{aligned}$$

and again by the fact that ϕ is one-to-one, it follows that

$$\phi^{-1}(ab) = \phi^{-1}(a) \cdot \phi^{-1}(b).$$

With respect to $\phi\psi$, to show that it is one-to-one, suppose that $\phi\psi(a) = \phi\psi(b)$. Since ϕ is one-to-one, we must have

$\psi(a) = \psi(b)$; since ψ is one-to-one, we must have $a = b$. This shows that $\phi\psi$ is one-to-one. That $\phi\psi$ is onto should be clear. The preservation of addition and multiplication by $\phi\psi$ follows as below:

$$\phi\psi(a+b) = \phi[\psi(a) + \psi(b)] = \phi\psi(a) + \phi\psi(b)$$

$$\phi\psi(ab) = \phi[\psi(a)\psi(b)] = \phi\psi(a) \cdot \phi\psi(b). \quad \blacksquare$$

A set G of automorphisms of **F** is called a *group of automorphisms* if it has the following three properties: (i) G contains the identity automorphism, (ii) G is closed under composition (i.e., $\phi, \psi \in G \Rightarrow \phi\psi \in G$), and (iii) G contains the inverse of every one of its elements. For our purposes, the word "group" will always refer to a group of automorphisms of a field. By the previous lemma and the remarks preceding it, the set of *all* automorphisms of a field **F** is a group; but there may be subsets of this set which are groups as well. For the field **C**, let id denote the identity mapping and ϕ the mapping given by complex conjugation. Then $\{\text{id}, \phi\}$ is a group, but $\{\phi\}$ is not.

If ϕ is an automorphism, a useful abbreviation is ϕ^n. When n is a positive integer ϕ^n stands for ϕ composed with itself n times; when n is a negative integer ϕ^n stands for ϕ^{-1} composed with itself $|n|$ times; ϕ^0 stands for the identity mapping. One reason why this abbreviation is so useful is that it obeys the usual laws of exponents: $\phi^{m+n} = \phi^m\phi^n$ and $(\phi^n)^k = \phi^{nk}$, which follow directly from the definition.

We shall only need to consider *finite groups*, that is, groups with a finite number of elements. In this case, we denote by $|G|$ the number of elements in G; it is called the *order* of G. Given a set of automorphisms, to decide whether it forms a group we might try to verify items (i), (ii), and (iii) in the definition of a group. But if the set is finite, then not all these conditions need to be checked, for we have:

LEMMA 29b. *If G is a finite set of automorphisms and if G is closed under composition, then G is a group.*

Proof. Given such a set G, let ϕ be one of its elements and let n be the total number of elements. Since G is closed under composition, the list $\phi, \phi^2, \phi^3, \ldots, \phi^{n+1}$ must be a subset of G and therefore must contain some element twice. Suppose $\phi^j = \phi^{j+k}$ for some $k \geqslant 1$. Since $\phi^{j+k} = \phi^j\phi^k$, we have $\phi^j = \phi^j\phi^k$. By the fact that ϕ^j is one-to-one, it follows that ϕ^k is the identity automorphism, id. To show that G contains the inverse of ϕ, we observe that if $k = 1$, ϕ is its own inverse, whereas if $k > 1$, ϕ^{k-1} is the inverse of ϕ. ∎

If H and G are groups of automorphisms of a field $\mathbf{F}$ and if $H \subset G$, then we say that H is a *subgroup* of G. For a very simple example, $H = \{\mathrm{id}\}$ is a subgroup of any group. What kinds of subsets of a group might be subgroups? An important step in answering this question for finite groups is the following necessary condition, called **Lagrange's Theorem**:

THEOREM 29. *If G is a finite group and if H is a subgroup of G, then $|H|$ is a divisor of $|G|$.*

Proof. Let us define $m = |H|$. Then we can represent the (distinct) elements of H by the list: $\psi_1, \psi_2, \ldots, \psi_m$. If this list includes all of G, we are done, for then $H = G$ and $|H| = |G|$. If there is some element ϕ_1 of G not in this list, we write down a second list below the first:

$$\begin{array}{ccccc} \psi_1 & \psi_2 & \psi_3 & \cdots & \psi_m \\ \phi_1\psi_1 & \phi_1\psi_2 & \phi_1\psi_3 & \cdots & \phi_1\psi_m. \end{array}$$

The elements in the second list are products of elements of G and hence are elements of G. They are all distinct, since if $\phi_1\psi_i = \phi_1\psi_j$, then $\phi_1^{-1}\phi_1\psi_i = \phi_1^{-1}\phi_1\psi_j$, and so $\psi_i = \psi_j$. Also the two lists have no elements in common, since if $\phi_1\psi_i = \psi_j$, then $\phi_1 = \psi_j\psi_i^{-1} \in H$, by closure, which contradicts the choice of ϕ_1. Now, if these two lists exhaust G, we are done, for then $|G| = 2|H|$. But if there exists some element ϕ_2 of G not in the list, we write a third list below the others:

$$\begin{array}{ccccc} \psi_1 & \psi_2 & \psi_3 & \cdots & \psi_m \\ \phi_1\psi_1 & \phi_1\psi_2 & \phi_1\psi_3 & \cdots & \phi_1\psi_m \\ \phi_2\psi_1 & \phi_2\psi_2 & \phi_2\psi_3 & \cdots & \phi_2\psi_m. \end{array}$$

By the previous argument the elements of the third row are distinct among themselves, and they are distinct from the elements of the first row. They are also distinct from the elements of the previous row, for if $\phi_2\psi_i = \phi_1\psi_j$, then $\phi_2 = \phi_1\psi_j\psi_i^{-1}$, which belongs to the second row, since $\psi_j\psi_i^{-1}$ is some element of H. But this would contradict the choice of ϕ_2. Continuing to adjoin new elements to the list in this way, m at a time, we must eventually exhaust G, since G is finite. If at that point we have k rows, then $mk = |G|$. Thus $|H|$ divides $|G|$. ∎

The set of elements in each row of the table used in the proof of Theorem 29 is called a *left coset of H*. The number of such cosets is given by $|G|/|H|$, which is called the *index of H in G*.

If G is a finite group and $\phi \in G$, it was shown in the proof of Lemma 29b that for some integer k, $\phi^k = \text{id}$. Letting k be the least such integer, look at the set of automorphisms $\{\phi, \phi^2, \ldots, \phi^k = \text{id}\}$. It is a group! For, $\phi^i\phi^j = \phi^{i+j} = \phi^r$, where r is the unique number between 1 and k, inclusive, such that $i + j \equiv r(\text{mod } k)$. This group is a subgroup of G, and we call it the *cyclic subgroup of G generated by ϕ*. If there is an element $\phi \in G$ such that the cyclic subgroup generated by ϕ is all of G, then we say that G is a *cyclic* group. Cyclic groups of automorphisms will be particularly important in our work.

It was noted earlier that automorphisms do not always commute. However, in certain special cases they do. A group G is said to be *Abelian* or *commutative* if every pair of its elements commute. We have the following simple result:

THEOREM 30. *If a group G is cyclic, then it is Abelian.*

Proof. By the definition of cyclic, G may be represented as $\{\phi, \phi^2, \ldots, \phi^n = \text{id}\}$, where $n = |G|$. By associativity we have $\phi^i\phi^j = \phi^{i+j} = \phi^j\phi^i$, so that any two elements commute. ∎

This section began with the assertion that in studying mathematical objects it often proves useful to study their symmetries, where a symmetry is a one-to-one mapping of the object onto itself which preserves the relevant structures. In the case of fields, we study groups of their automorphisms. Much can be learned about the structure of the fields in question by studying the structure of certain automorphism groups. By the 'structure' of groups, we mean things like how many and what kind of subgroups they have, whether they are cyclic or not, and similar questions.

It is worth remarking at this point that all the properties of automorphism groups developed in this section could have been defined or derived for any mathematical system having the same basic properties as such groups. This more general point of view may be found in textbooks on abstract algebra.

PROBLEMS

1. Suppose ϕ is a mapping from a field $\mathbf{F}$ to itself such that ϕ is one-to-one, onto, and preserves addition and multiplication (i.e., $\phi(a+b)=\phi(a)+\phi(b)$ and $\phi(ab) = \phi(a)\phi(b)$ for all $a, b \in \mathbf{F}$). Show that $\phi(0)=0$, $\phi(1)=1$, and that ϕ is an automorphism of $\mathbf{F}$.

2. Let ϕ be an automorphism of a field of $\mathbf{F}$. Show that for every rational number q, $\phi(q)=q$. (That is, ϕ, restricted to the subfield $\mathbf{Q}$ of $\mathbf{F}$, is the identity mapping.)

3. Find all the automorphisms of $\mathbf{Q}(\sqrt[3]{2})$.

4. Find all the automorphisms of $\mathbf{Q}(\sqrt{2})$.

5. Prove Theorem 29 using right cosets. (If H is a subgroup of G, a right coset of H is a set $\{\psi_i\phi\}$ where the ψ_i's range through H and ϕ is a fixed element of G.)

6. Let G be any group, finite or not, and let H be a subgroup of G. If $\phi \in G$, then the set $\phi H = \{\phi\psi \mid \psi \in H\}$ is called the left coset of H by ϕ. Show that any two left cosets of G are either identical or disjoint.

7. If G is a group whose order is a prime p, how many subgroups does G have?

8. If G is a cyclic group of order n, determine exactly how many elements of G generate G.

9. If G is a cyclic group of order n, and if $m \mid n$, show that G has a cyclic subgroup of order m.

10. Show that every subgroup of a cyclic group is cyclic.

11. If $|G| = p$, a prime, show that G is cyclic.

12. If H is a subgroup of prime index p in G, show that other than for H and G, there is no subgroup $\tilde{H}$ such that $H \subset \tilde{H} \subset G$.

Section 3.3. The Group of an Extension

In this section we shall introduce those groups of automorphisms that will be of use in our work. If $\mathbf{E}$ is an extension of the field $\mathbf{F}$, consider the set of all automorphisms of $\mathbf{E}$ which leave each element of $\mathbf{F}$ unchanged. This set is readily seen to be a group. First, it obviously contains the identity automorphism on $\mathbf{E}$. Second, it contains the inverse of each member; for $\phi(a) = a$ if and only if $a = \phi^{-1}(a)$ and so the automorphisms ϕ and ϕ^{-1} leave the same elements unchanged. Third, it is closed under composition; if the automorphisms ϕ and ψ leave every $a \in \mathbf{F}$ unchanged, then so too does the automorphism $\psi\phi$, for $\psi\phi(a) = \psi(\phi(a)) = \psi(a) = a$. This group of automorphisms is called the *group of* $\mathbf{E}$ *over* $\mathbf{F}$, and it is denoted $G(\mathbf{E}/\mathbf{F})$.

How many such automorphisms might there be? An infinite number? A finite number? What do they look like? Well, recalling the results of Problems 3 and 4 of the previous section, there is only one element of $G(\mathbf{Q}(\sqrt[3]{2})/\mathbf{Q})$ and there are two elements of $G(\mathbf{Q}(\sqrt{2})/\mathbf{Q})$. In order to answer these questions in general, we first need some terminology. If a number r is algebraic over $\mathbf{F}$, then r is the root of some irreducible polynomial over **F**, and this polynomial is unique up to a constant factor (Section 2.2). We denote the complete set of roots of this polynomial $r = r_1, r_2, \ldots, r_n$, where n is its degree. These roots are called *conjugates* of each other. They are distinct, since an irreducible polynomial cannot have multiple roots (Problem 10, Section 2.1). The simple algebraic extension $\mathbf{F}(r)$ may or may not contain any conjugates of r (other than r itself). The following theorem and its proof yield considerable information about the structure of the elements of $G(\mathbf{E}/\mathbf{F})$ when **E** is written as a simple extension of **F**.

THEOREM 31. *If* $\mathbf{E} = \mathbf{F}(r)$, *where* r *is algebraic over* **F** *with conjugates* $r = r_1, r_2, \ldots, r_n$, *then for each* $\phi \in G(\mathbf{E}/\mathbf{F})$, $\phi(r) =$ *some* r_i; *moreover, for each* $r_i \in \mathbf{E}$, *there is a unique* $\phi \in G(\mathbf{E}/\mathbf{F})$ *satisfying* $\phi(r) = r_i$. *If* **E** *is any finite extension of* **F**, *then* $|G(\mathbf{E}/\mathbf{F})| \leqslant [\mathbf{E} : \mathbf{F}]$.

Proof. The second sentence will follow immediately from the first, since by Theorem 18, every finite extension is a simple algebraic extension, $\mathbf{E} = \mathbf{F}(r)$, and the total number n of conjugates of r satisfies $n = \deg_{\mathbf{F}} r = [\mathbf{E} : \mathbf{F}]$; so certainly there are *at most n* conjugates *in* **E**.

Let $\phi \in G(\mathbf{E}/\mathbf{F})$. We shall make use of the fact that if $p(x) = c_0 + c_1x + c_2x^2 + \cdots + c_kx^k$ is any polynomial with coefficients in **F**, then $\phi(p(r)) = p(\phi(r))$. To see this, observe that $\phi(p(r)) = \phi(c_0) + \phi(c_1)[\phi(r)] + \phi(c_2)[\phi(r)]^2 + \cdots + \phi(c_k)[\phi(r)]^k$ by the fact that ϕ is an automorphism, and then from this that $\phi(p(r)) = c_0 + c_1[\phi(r)] + c_2[\phi(r)]^2 + \cdots + c_k[\phi(r)]^k = p(\phi(r))$ by the fact that ϕ leaves the elements of **F** unchanged.

Now, if f is the minimal polynomial of r over **F** and if $\phi \in G(\mathbf{E}/\mathbf{F})$, then $0 = \phi(0) = \phi(f(r)) = f(\phi(r))$ and so $\phi(r)$ must be some conjugate r_i of r. Conversely, if $r_i \in \mathbf{E}$ we want to show that there is a unique $\phi \in G(\mathbf{E}/\mathbf{F})$ such that $\phi(r) = r_i$. This is more complicated.

By Theorem 17, any element $a \in \mathbf{E}$ can be expressed in the form $a = p(r)$ for some polynomial $p \in \mathbf{F}[x]$. (If we restrict the degree of p to be $\leqslant n - 1$, then this representation is unique, but this restriction is not convenient for our present purposes.) From the equation $\phi(a) = \phi(p(r)) = p(\phi(r))$ it follows that an element $\phi \in G(\mathbf{E}/\mathbf{F})$ is uniquely determined by its value $\phi(r)$. Thus there is at most one $\phi \in G(\mathbf{E}/\mathbf{F})$ such that $\phi(r) = r_i$.

In accordance with this computation, let us try to construct such a ϕ by the natural rule. That is, for each $a \in \mathbf{E}$, write $a = p(r)$ for an appropriate $p \in \mathbf{F}[x]$ and then define $\phi(a) = p(r_i)$. To see that this mapping is well defined, suppose a has two representations: $a = p(r) = \tilde{p}(r)$, where both p and $\tilde{p}$ belong to $\mathbf{F}[x]$. We want to verify that $p(r_i) = \tilde{p}(r_i)$. This follows from the fact that the polynomial $p - \tilde{p}$, which has r as a root, must therefore be divisible by the minimal polynomial f of r over $\mathbf{F}$ (Problem 6, Section 2.1). Since r_i is also a root of f, $p - \tilde{p}$ has r_i for a root. That is, $p(r_i) = \tilde{p}(r_i)$.

This mapping ϕ leaves elements of $\mathbf{F}$ unchanged, for such elements may be represented by constant polynomials. That ϕ is onto follows from the fact (Problem 13, Section 2.2) that $\mathbf{E} = \mathbf{F}(r_i)$. In particular, any element of $\mathbf{E}$ can be represented in the form $p(r_i)$ for some $p \in \mathbf{F}[x]$, and $p(r_i)$ is the image under ϕ of $p(r)$, which is in $\mathbf{E}$. To see that ϕ is one-to-one, suppose that $\phi(q(r)) = \phi(\tilde{q}(r))$, where q and $\tilde{q}$ are polynomials in $\mathbf{F}[x]$. By the definition of ϕ, $q(r_i) = \tilde{q}(r_i)$ and so, by reasoning exactly analogous to that in the previous paragraph, $q(r) = \tilde{q}(r)$. Thus ϕ is one-to-one. It remains to show that ϕ is an automorphism, and by Problem 1 of Section 3.2, for this it suffices to show that it preserves addition and multiplication.

Let a and b be arbitrary elements of $\mathbf{E}$, represented respectively, as $p(r)$ and $q(r)$, where p and $q \in \mathbf{F}[x]$. Letting $t \in \mathbf{F}[x]$ be the product of the polynomials p and q, we have:

$$\phi(ab) = \phi(t(r)) = t(r_i) = p(r_i)q(r_i) = \phi(a)\phi(b).$$

Letting $s \in \mathbf{F}[x]$ be the sum of p and q,

$$\phi(a + b) = \phi(s(r)) = s(r_i) = p(r_i) + q(r_i) = \phi(a) + \phi(b).$$

Thus the mapping induced by the condition $\phi(r) = r_i$ is an automorphism. This completes the proof. ∎

Returning now to the examples preceding Theorem 31, $G(\mathbf{Q}(\sqrt[3]{2})/\mathbf{Q})$ and $G(\mathbf{Q}(\sqrt{2})/\mathbf{Q})$, of respective orders one and two, we note that in the first case the single automorphism not only leaves each element of $\mathbf{Q}$ unchanged, it leaves other things unchanged as well, namely all of $\mathbf{Q}(\sqrt[3]{2})$. The second case is different, however, in that the set of elements left unchanged by every (i.e., both) element of the group is precisely $\mathbf{Q}$ itself. This is an important distinction.

If ϕ is an automorphism of a field $\mathbf{E}$, then the set $\mathbf{K}_\phi = \{a \in \mathbf{E} \mid \phi(a) = a\}$ is readily seen to be a field itself. For example, to verify closure under addition, let $a, b \in \mathbf{K}_\phi$. Then $\phi(a + b) = \phi(a) + \phi(b) = a + b$. Consequently $a + b \in \mathbf{K}_\phi$. The other operations are handled in the same way. By Problem 2 of Section 3.2, $\mathbf{K}_\phi$ contains all the rational numbers. The field $\mathbf{K}_\phi$ is called the *fixed field* of ϕ. If G is any set of automorphisms of $\mathbf{E}$, then the intersection of their corresponding fixed fields, $\mathbf{K} = \bigcap_{\phi \in G} \mathbf{K}_\phi$, is called the *fixed field of G*. It is, of course, a field since the intersection of an arbitrary collection of fields is always a field. We have just had one example where the fixed field of $G(\mathbf{E}/\mathbf{F})$ is $\mathbf{E}$ and another example where the fixed field of $G(\mathbf{E}/\mathbf{F})$ is $\mathbf{F}$. Obviously, the fixed field **K** of $G(\mathbf{E}/\mathbf{F})$ is always an intermediate field between **E** and **F**; that is, $\mathbf{E} \supset \mathbf{K} \supset \mathbf{F}$.

In the light of the preceding developments, there are certain natural questions one might ask. Under what conditions are the above relations a little 'tighter'? For example, when do we have $|G(\mathbf{E}/\mathbf{F})| = [\mathbf{E} : \mathbf{F}]$? When is the fixed field of $G(\mathbf{E}/\mathbf{F})$ simply **F** itself? Not only might these questions appear inherently interesting, but in fact they are very close to the core of the present investigation. Definitive answers will be given in the next section, but the interested reader may prefer to think about them independently first.

If f is a polynomial over **F** and if **E** is its splitting field over **F**, then $G(\mathbf{E}/\mathbf{F})$ is called the *Galois group* of f over **F**. If the field **F** is not specified, it is understood to be the smallest field containing all the coefficients of f. The Galois group of a polynomial is the group whose structure will ultimately be of interest and for whose study we need to develop the more general theory of the groups of field extensions, as in this and the following sections. Determining the Galois group of a polynomial can be quite difficult, but fortunately we shall not have need of a general procedure for this. We have looked at one very simple case already, of course, namely $f(x) = x^2 - 2$. For the splitting field is $\mathbf{Q}(\sqrt{2})$ and $G(\mathbf{Q}(\sqrt{2})/\mathbf{Q})$ has been shown to have two elements, the identity mapping and the mapping $\phi : a + b\sqrt{2} \to a - b\sqrt{2}$, $a, b \in \mathbf{R}$. Other examples will appear in the problems and elsewhere. (Problem 1 will be very important for later work.)

PROBLEMS

1. If p is a prime, show that the Galois group of $x^{p-1} + x^{p-2} + \cdots + 1 = (x^p - 1)/(x - 1)$ is cyclic and of order $p - 1$.

2. Determine the order of $G(\mathbf{E}/\mathbf{Q})$ where $\mathbf{E} = \mathbf{Q}(\sqrt{2}, \sqrt{3})$. (Hint: See Problem 7, Section 2.3, and Problems 1 and 2, Section 1.4.)

3. What is the order of the Galois group of $x^5 - 1$?

4. Is there some polynomial $f \in \mathbf{Q}[x]$ such that $\mathbf{Q}(\sqrt[3]{2})$ is its splitting field?

5. If $\phi \in G(\mathbf{E}/\mathbf{F})$, where $\mathbf{E}$ is a finite extension of $\mathbf{F}$, and if $\mathbf{K}$ is an intermediate field, show that the set $\phi(\mathbf{K}) = \{\phi(a) \mid a \in \mathbf{K}\}$ is also an intermediate field. Show further that $[\mathbf{K} : \mathbf{F}] = [\phi(\mathbf{K}) : \mathbf{F}]$.

6. Let $\mathbf{F}$ be a field. Numbers $a_1, a_2, \ldots, a_n$ are said to be *algebraically independent* over $\mathbf{F}$ if there is no nontrivial polynomial $p(x_1, x_2, \ldots, x_n)$ over $\mathbf{F}$ such that $p(a_1, a_2, \ldots, a_n) = 0$. Show that if $\mathbf{F}$ is countable, then for any n there exist n algebraically independent elements over $\mathbf{F}$.

7. Let $a_1, a_2, \ldots, a_n$ be algebraically independent over a field $\mathbf{F}$. Define $\mathbf{E} = \mathbf{F}(a_1, a_2, \ldots, a_n)$. Show that for every permutation of the numbers $a_1, a_2, \ldots, a_n$ there exists a unique $\phi \in G(\mathbf{E}/\mathbf{F})$ which has the effect of this permutation on the a_i's.

8. How many automorphisms of $\mathbf{R}$ are there?

*9. If the field of complex numbers $\mathbf{C}$ is an algebraic extension of a field $\mathbf{E}$, show that any automorphism of $\mathbf{E}$ can be extended to an automorphism of $\mathbf{C}$.

*10. How many automorphisms of $\mathbf{C}$ are there?

Section 3.4. Two Fundamental Theorems

If $\mathbf{E}$ is a finite extension of $\mathbf{F}$ with the property that every irreducible polynomial over $\mathbf{F}$ that has one root in $\mathbf{E}$ has all its roots in $\mathbf{E}$, then $\mathbf{E}$ is said to be a *normal extension* of $\mathbf{F}$. At first, this may be a difficult property to appreciate; it is perhaps best clarified by moving directly to the following major theorem:

THEOREM 32. *Let* **E** *be a finite extension of* **F**. *Then the following statements are equivalent*:

(i) **E** *is a normal extension of* **F**.
(ii) **E** *is the splitting field of a polynomial over* **F**.
(iii) $|G(\mathbf{E}/\mathbf{F})| = [\mathbf{E} : \mathbf{F}]$.
(iv) **F** *is the fixed field of* $G(\mathbf{E}/\mathbf{F})$.
(v) **E** *is a normal extension of every intermediate field* $\mathbf{K}$, $\mathbf{E} \supset \mathbf{K} \supset \mathbf{F}$.

Proof. In order to show the equivalence of the above statements, the seven implications represented by arrows in the following diagram will be established:

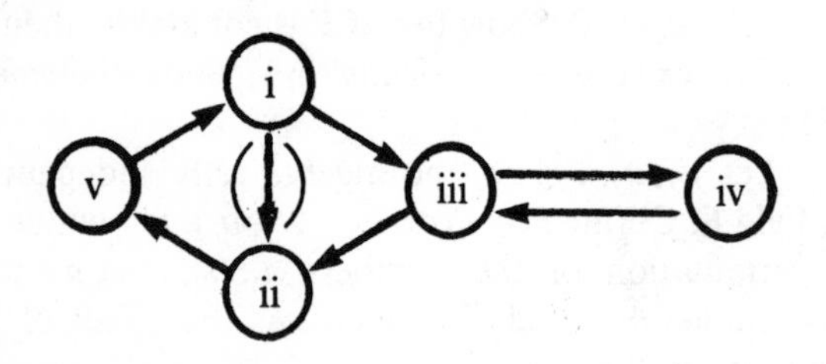

The implication (i$\Rightarrow$ii) is not logically necessary to the proof, but as it is very brief and helps to clarify the relation between the statements, it will be included.

(i$\Rightarrow$ii) Since **E** is a finite extension of **F**, by Theorem 18, **E** is a simple algebraic extension, $\mathbf{E} = \mathbf{F}(r)$. If f is the minimal polynomial for r over **F**, **E** is the splitting field of f. For, if $r = r_1, r_2, \ldots, r_n$ are the conjugates of r, $\mathbf{E} = \mathbf{F}(r) \subset \mathbf{F}(r_1, r_2, \ldots, r_n) \subset \mathbf{E}$, the last inclusion following from normality.

(ii$\Rightarrow$v) Let **E** be the splitting field of a polynomial f over **F** with roots $r_1, r_2, \ldots, r_n$. Then $\mathbf{E} = \mathbf{F}(r_1, r_2, \ldots, r_n)$

and we may build up to **E** by a sequence of simple extensions: $\mathbf{F}_1 = \mathbf{F}(r_1)$, $\mathbf{F}_2 = \mathbf{F}_1(r_2)$, $\mathbf{F}_3 = \mathbf{F}_2(r_3), \ldots, \mathbf{F}_n = \mathbf{F}_{n-1}(r_n)$. A basis for the extension $\mathbf{F}_j$ over $\mathbf{F}_{j-1}$ was shown in the proof of Theorem 17 to be $1, r_j, r_j^2, \ldots, r_j^{m-1}$, where $m = \deg_{\mathbf{F}_{j-1}}(r_j)$, which may be anything from 1 to n. As in the proof of Theorem 19, a basis for the extension **E** over **F** may be obtained by forming all products such that one factor is chosen from the basis of each of these single simple extensions. We conclude that each element $a \in \mathbf{E}$ may be written in the form $a = P(r_1, r_2, \ldots, r_n)$, where P is a polynomial over **F** in n variables.

Now we apply these facts to show that if g is an irreducible polynomial over a field **K**, where $\mathbf{F} \subset \mathbf{K} \subset \mathbf{E}$, and if g has one root $a \in \mathbf{E}$, then all its roots must be in **E**. By the previous paragraph, we may write $a = P(r_1, r_2, \ldots, r_n)$, where P is a polynomial over **F** in n variables. Now look at all the values of P under all permutations of r_i's; that is, if π_i is one of the $n!$ permutations on the numbers 1 through n, define $a_i = P(r_{\pi_i(1)}, r_{\pi_i(2)}, \ldots, r_{\pi_i(n)})$. Taking π_1 as the identity permutation, the resulting values are $a = a_1, a_2, \ldots, a_{n!}$, each of which is in **E**. We will show that g divides the polynomial h defined by $h(x) = \prod_{i=1}^{n!}(x - a_i)$, and hence that all its roots are included among those of h, all of which are in **E**.

To show that g divides h, we first show that h has all of its coefficients in **F**. Letting σ_i denote the ith elementary symmetric function of $n!$ variables and letting $P_{\pi_i}(x_1, x_2, \ldots, x_n) = P(x_{\pi_i(1)}, x_{\pi_i(2)}, \ldots, x_{\pi_i(n)})$, we have that the coefficient of $x^{n!-i}$ in h is simply $\pm\sigma_i(P_{\pi_1}, P_{\pi_2}, \ldots, P_{\pi_{n!}})$ evaluated at $(r_1, r_2, \ldots, r_n)$. However the polynomial function of n variables given by $\sigma_i(P_{\pi_1}, P_{\pi_2}, \ldots, P_{\pi_{n!}})$ is a symmetric polynomial over **F**, for any permutation of the variables simply permutes the

P's, under which σ_i is invariant. Therefore, by Corollary 9a the value of $\sigma_i(P_{\pi_1}, P_{\pi_2}, \ldots, P_{\pi_{n!}})$ when evaluated at the roots $(r_1, r_2, \ldots, r_n)$ of a polynomial over $\mathbf{F}$, is in $\mathbf{F}$. Since we now have $h \in \mathbf{F}[x]$, certainly $h \in \mathbf{K}[x]$. Now both g and $h \in \mathbf{K}[x]$, they have one root a in common, and g is irreducible. It follows (Cf. Problem 6, Section 2.1) that g divides h in $\mathbf{K}[x]$. As noted at the outset, this implies that all of the roots of g, being among those of h, are in $\mathbf{E}$. Thus $\mathbf{E}$ is a normal extension of $\mathbf{K}$.

(v$\Rightarrow$i) This is obvious, as $\mathbf{F}$ itself is an intermediate field.

(i$\Rightarrow$iii) Since $\mathbf{E}$ is a finite extension of $\mathbf{F}$, we know we can write $\mathbf{E} = \mathbf{F}(r)$, where r has conjugates $r_1, r_2, \ldots, r_n$. Also, we must have $n = [\mathbf{E} : \mathbf{F}]$. By normality, each $r_i \in \mathbf{E}$. By Theorem 31, corresponding to each r_i there is exactly one element of $G(\mathbf{E}/\mathbf{F})$, and there are no other elements of $G(\mathbf{E}/\mathbf{F})$. Thus $|G(\mathbf{E}/\mathbf{F})| = [\mathbf{E} : \mathbf{F}]$.

(iii$\Rightarrow$ii) As in the previous paragraph, r must have the same number of conjugates in $\mathbf{E}$ as there are automorphisms in $G(\mathbf{E}/\mathbf{F})$. So if $|G(\mathbf{E}/\mathbf{F})| = [\mathbf{E} : \mathbf{F}] = n$, all n conjugates of r must be in $\mathbf{E}$. Thus $\mathbf{E}$ is the splitting field for the minimal polynomial of r over $\mathbf{F}$.

(iii$\Rightarrow$iv) Let $\mathbf{K}$ be the fixed field of $G(\mathbf{E}/\mathbf{F})$. Clearly $\mathbf{E} \supset \mathbf{K} \supset \mathbf{F}$. By (v), which has been shown to be equivalent to (iii), $\mathbf{E}$ is a normal extension of $\mathbf{K}$. Since i$\Rightarrow$iii, $|G(\mathbf{E}/\mathbf{K})| = [\mathbf{E} : \mathbf{K}]$. But $G(\mathbf{E}/\mathbf{K}) = G(\mathbf{E}/\mathbf{F})$, since every automorphism of $\mathbf{E}$ which leaves $\mathbf{F}$ fixed also leaves $\mathbf{K}$ fixed. Thus we have: $[\mathbf{E} : \mathbf{K}] = |G(\mathbf{E}/\mathbf{K})| = |G(\mathbf{E}/\mathbf{F})| = [\mathbf{E} : \mathbf{F}]$. By Theorem 19, $[\mathbf{E} : \mathbf{F}] = [\mathbf{E} : \mathbf{K}][\mathbf{K} : \mathbf{F}]$; it follows that $[\mathbf{K} : \mathbf{F}] = 1$; that is, $\mathbf{K} = \mathbf{F}$.

(iv$\Rightarrow$iii) Again we begin by writing $\mathbf{E} = \mathbf{F}(r)$. Let $n = [\mathbf{E} : \mathbf{F}] = \deg_{\mathbf{F}} r$, so that any polynomial over $\mathbf{F}$ of which r is a root must have degree $\geqslant n$. Letting $m = |G(\mathbf{E}/\mathbf{F})|$, we will construct a polynomial of degree m

over **F** having r as a root, from which we will thus conclude $m \geqslant n$. By Theorem 31, then, $m = n$.

Let $\phi_1, \phi_2, \ldots, \phi_m$ be the elements of $G(\mathbf{E}/\mathbf{F})$. We will show that the polynomial $h(x) = \prod_{i=1}^{m}(x - \phi_i(r))$ has all its coefficients in **F** by showing that each coefficient is unchanged by every $\phi \in G(\mathbf{E}/\mathbf{F})$. The coefficient of x^{m-i} is $\pm\sigma_i(\phi_1(r), \phi_2(r), \ldots, \phi_m(r))$, where σ_i is the ith elementary symmetric function in m variables, a symmetric polynomial over **Q** and hence over **F**. Letting ϕ be an arbitrary automorphism in $G(\mathbf{E}/\mathbf{F})$, $\phi[\sigma_i(\phi_1(r), \phi_2(r), \ldots, \phi_m(r))] = \sigma_i(\phi\phi_1(r), \phi\phi_2(r), \ldots, \phi\phi_m(r)) = \sigma_i(\phi_1(r), \phi_2(r), \ldots, \phi_m(r))$ since σ_i is symmetric and the list $\phi\phi_1, \phi\phi_2, \ldots, \phi\phi_m$, representing all of $G(\mathbf{E}/\mathbf{F})$, must be a permutation of the list $\phi_1, \phi_2, \ldots, \phi_m$. Since ϕ is arbitrary, each coefficient in h, being unchanged by every element of $G(\mathbf{E}/\mathbf{F})$, must be in **F**. But r is a root of this polynomial, since the identity automorphism must be in $G(\mathbf{E}/\mathbf{F})$, and so $m \geqslant n$. As noted earlier, this suffices to complete the proof. ∎

As a corollary to this theorem, we obtain a natural one-to-one correspondence between intermediate fields and subgroups of $G(\mathbf{E}/\mathbf{F})$, called the **fundamental Galois pairing**:

COROLLARY 32. *If* **E** *is a normal extension of* **F**, *there is a one-to-one correspondence between the intermediate fields* **K**, $\mathbf{E} \supset \mathbf{K} \supset \mathbf{F}$, *and the subgroups of* $G(\mathbf{E}/\mathbf{F})$; *this correspondence associates* **K** *with* $G(\mathbf{E}/\mathbf{K})$. *The correspondence reverses inclusion relations*: $\mathbf{K}_1 \subset \mathbf{K}_2$ *if and only if* $G(\mathbf{E}/\mathbf{K}_1) \supset G(\mathbf{E}/\mathbf{K}_2)$.

Proof. With each intermediate field **K** we associate the group $G(\mathbf{E}/\mathbf{K})$; **K** is the fixed field of this group since by

Theorem 32 $\mathbf{E}$ is normal over $\mathbf{K}$. Every subgroup of $G(\mathbf{E}/\mathbf{F})$ is associated with some such field, namely its fixed field. If $G(\mathbf{E}/\mathbf{K}_1) = G(\mathbf{E}/\mathbf{K}_2)$, then the two groups have the same fixed field. By normality of $\mathbf{E}$ over both $\mathbf{K}_1$ and $\mathbf{K}_2$, $\mathbf{K}_1 = \mathbf{K}_2$. It is obvious by the definitions of the group of an extension and of the fixed field that inclusion relations are reversed by this correspondence. ∎

Note that in the absence of the assumption that $\mathbf{E}$ is a normal extension of $\mathbf{F}$, the correspondence given here would no longer be one-to-one, for several fields might have the same group associated with them. For example, if $\mathbf{E} = \mathbf{Q}(\sqrt[3]{2})$, then both $\mathbf{E}$ and $\mathbf{Q}$ have the same group associated with them, namely the group consisting only of the identity automorphism on $\mathbf{E}$.

From Theorem 32 and its corollary, we can begin to get some feeling for the properties of normal extensions and for the correspondence between fields and groups. However, no mention is made in these statements about the question: If $\mathbf{E}$ is a normal extension of $\mathbf{F}$, under what circumstances is an intermediate field $\mathbf{K}$ also a normal extension of $\mathbf{F}$? While we know that $\mathbf{E}$ must be a normal extension of $\mathbf{K}$, certainly $\mathbf{K}$ need not be a normal extension of $\mathbf{F}$. For example, if $\mathbf{E}$ is the splitting field of $x^3 - 2$ over $\mathbf{Q}$, one intermediate field is $\mathbf{Q}(\sqrt[3]{2})$, and we have already seen (Problem 4, Section 3.3) that this is not a splitting field of any polynomial over $\mathbf{Q}$, and hence by Theorem 32 it cannot be a normal extension of $\mathbf{Q}$. An alternate way to conclude that $\mathbf{Q}(\sqrt[3]{2})$ is not normal over $\mathbf{Q}$ is to observe that $|G(\mathbf{Q}(\sqrt[3]{2})/\mathbf{Q})| = 1 \neq 3 = [\mathbf{Q}(\sqrt[3]{2}) : \mathbf{Q}]$.

Those intermediate fields $\mathbf{K}$, $\mathbf{E} \supset \mathbf{K} \supset \mathbf{F}$, such that $\mathbf{K}$ is normal over $\mathbf{F}$ are extremely important in the problem of solvability by radicals. It will be seen below that such fields correspond to a special kind of subgroup of $G(\mathbf{E}/\mathbf{F})$.

If H is a subgroup of a group G, and if $\phi \in G$, let us define the set $\phi^{-1}H\phi = \{\phi^{-1}\psi\phi \mid \psi \in H\}$. It is a subgroup of G. For, since $\text{id} \in H$, $\phi^{-1}\text{id}\phi = \phi^{-1}\phi = \text{id} \in \phi^{-1}H\phi$; since $\psi^{-1} \in H$, $(\phi^{-1}\psi\phi)^{-1} = \phi^{-1}\psi^{-1}\phi \in \phi^{-1}H\phi$; and if $\psi, \tau \in H$, $(\phi^{-1}\psi\phi)(\phi^{-1}\tau\phi) = \phi^{-1}(\psi\tau)\phi \in \phi^{-1}H\phi$. Also, since G is closed, $\phi^{-1}H\phi \subset G$. The subgroup $\phi^{-1}H\phi$ is said to be a *conjugate* subgroup of H. Two subgroups H_1 and H_2 of G are said to be conjugate if $H_1 = \phi^{-1}H_2\phi$ for some $\phi \in G$. The relation of conjugacy is easily seen to be an equivalence relation on the set of subgroups of G; it is reflexive, symmetric, and transitive. (See Problem 3.) A subgroup H of G is said to be a *normal subgroup*, or an *invariant subgroup*, if $\phi^{-1}H\phi = H$ for every $\phi \in G$; that is, if H equals each of its conjugates.

Two extensions $\mathbf{K}_1$ and $\mathbf{K}_2$ of a field $\mathbf{F}$ are said to be *conjugate* over $\mathbf{F}$ if $\mathbf{K}_1 = \mathbf{F}(r_1)$ and $\mathbf{K}_2 = \mathbf{F}(r_2)$, where r_1 and r_2 are conjugates over $\mathbf{F}$, that is, r_1 and r_2 have the same minimal polynomial over $\mathbf{F}$. Under the fundamental Galois pairing, conjugate fields will be shown to correspond to conjugate subgroups, and this fact, called the Fundamental Theorem of Galois Theory, is at the core of the solution to the problem of solvability by radicals. It is also of intrinsic interest because of the structural questions it clarifies. But in order to deal conveniently with conjugate fields, we first need the following preliminary result:

LEMMA 33. *Let* $\mathbf{E}$ *be a normal extension of* $\mathbf{F}$. *Intermediate fields* $\mathbf{K}_1$ *and* $\mathbf{K}_2$ *are conjugate if and only if there exists some* $\phi \in G(\mathbf{E}/\mathbf{F})$ *such that* $\phi(\mathbf{K}_1) = \mathbf{K}_2$.

Proof. (if) Assume that there is such a $\phi \in G(\mathbf{E}/\mathbf{F})$ and write $\mathbf{K}_1$ as a simple algebraic extension of $\mathbf{F}$,

$\mathbf{K}_1 = \mathbf{F}(r_1)$. Then $\phi(r_1)$ must be a conjugate of r_1, for if f is the minimal polynomial of r_1 over $\mathbf{F}$, $0 = \phi(f(r_1)) = f(\phi(r_1))$. Let $r_2 = \phi(r_1) \in \mathbf{K}_2$. Clearly $\mathbf{F}(r_2) \subset \mathbf{K}_2$. To show equality, we have $[\mathbf{K}_2 : \mathbf{F}] = [\mathbf{K}_1 : \mathbf{F}] = \deg_{\mathbf{F}} r_1 = \deg_{\mathbf{F}} r_2 = [\mathbf{F}(r_2) : \mathbf{F}]$, where the first equality follows from Problem 5 of Section 3.3. But since it is also true that $[\mathbf{K}_2 : \mathbf{F}] = [\mathbf{K}_2 : \mathbf{F}(r_2)][\mathbf{F}(r_2) : \mathbf{F}]$, we conclude that $[\mathbf{K}_2 : \mathbf{F}(r_2)] = 1$. That is, $\mathbf{F}(r_2) = \mathbf{K}$. Hence $\mathbf{K}_1$ and $\mathbf{K}_2$ are conjugate fields.

(only if) If $\mathbf{K}_1 = \mathbf{F}(r_1)$ and $\mathbf{K}_2 = \mathbf{F}(r_2)$, where r_1 and r_2 are conjugates of each other, we want to show that there exists a $\phi \in G(\mathbf{E}/\mathbf{F})$ such that $\phi(\mathbf{K}_1) = \mathbf{K}_2$. Suppose $[\mathbf{K}_1 : \mathbf{F}] = n$, so there are n conjugates $r_1, r_2, \ldots, r_n$. It is actually easier to establish the stronger result that for each r_j there are exactly $[\mathbf{E} : \mathbf{K}_1]$ elements ϕ of $G(\mathbf{E}/\mathbf{F})$ such that $\phi(r_1) = r_j$. Let r_j be fixed and such that there is at least one $\psi \in G(\mathbf{E}/\mathbf{F})$ for which $\psi(r_1) = r_j$. (Initially, we do not know whether any such r_j's exist.) Let $S_j = \{\phi \in G(\mathbf{E}/\mathbf{F}) \mid \phi(r_1) = r_j\}$ and observe that the set $\psi^{-1}S_j = \{\psi^{-1}\phi \mid \phi \in S_j\}$ is just $G(\mathbf{E}/\mathbf{K}_1)$. For, each element of $\psi^{-1}S$ leaves $\mathbf{K}_1$ unchanged, and any $\phi \in G(\mathbf{E}/\mathbf{K}_1)$ can be written $\phi = \psi^{-1}(\psi\phi)$, in which it is clear that $\psi\phi \in S_j$. Thus S_j is the left coset $\psi G(\mathbf{E}/\mathbf{K}_1)$. If m is the number of conjugates of r_1 for which there are such nonempty sets S_j, each of which is now known to have exactly $|G(\mathbf{E}/\mathbf{K}_1)|$ elements, we have (since every ϕ is in some S_j)

$$|G(\mathbf{E}/\mathbf{F})| = m|G(\mathbf{E}/\mathbf{K}_1)|.$$

By the normality of $\mathbf{E}$ over $\mathbf{F}$ and over $\mathbf{K}_1$,

$$[\mathbf{E} : \mathbf{F}] = m[\mathbf{E} : \mathbf{K}_1],$$

from which it follows that $m = [\mathbf{K}_1 : \mathbf{F}] = n$.

By the result of the previous paragraph, $\phi(\mathbf{K}_1) \supset \mathbf{F}(r_2)$; but since every element of $\phi(\mathbf{K}_1)$ is a polynomial over $\mathbf{F}$ evaluated at r_2, in fact $\phi(\mathbf{K}_1) = \mathbf{F}(r_2) = \mathbf{K}_2$. ∎

And now for the **Fundamental Theorem of Galois Theory:**

THEOREM 33. *Let* $\mathbf{E}$ *be a normal extension of* $\mathbf{F}$. *Two intermediate fields* $\mathbf{K}_1$ *and* $\mathbf{K}_2$ *are conjugate over* $\mathbf{F}$ *if and only if* $G(\mathbf{E}/\mathbf{K}_1)$ *and* $G(\mathbf{E}/\mathbf{K}_2)$ *are conjugate subgroups of* $G(\mathbf{E}/\mathbf{F})$. $\mathbf{K}$ *is a normal extension of* $\mathbf{F}$ *if and only if* $G(\mathbf{E}/\mathbf{K})$ *is a normal subgroup of* $G(\mathbf{E}/\mathbf{F})$.

Proof. If $\mathbf{K}_1$ and $\mathbf{K}_2$ are conjugate over $\mathbf{F}$, then by the preceding lemma there exists an automorphism $\phi \in G(\mathbf{E}/\mathbf{F})$ such that $\phi(\mathbf{K}_1) = \mathbf{K}_2$. Using the abbreviation $H = G(\mathbf{E}/\mathbf{K}_2)$, it will follow easily that $G(\mathbf{E}/\mathbf{K}_1) = \phi^{-1}H\phi$. For if $\psi \in G(\mathbf{E}/\mathbf{K}_1)$, then $\psi = \phi^{-1}(\phi\psi\phi^{-1})\phi$ and for every $a \in \mathbf{K}_2$, $\phi\psi\phi^{-1}(a) = \phi\psi[\phi^{-1}(a)] = \phi\phi^{-1}(a) = a$, since ψ leaves $\phi^{-1}(a) \in \mathbf{K}_1$ fixed. Therefore $\phi\psi\phi^{-1} \in G(\mathbf{E}/\mathbf{K}_2)$. Conversely, if $\psi \in H = G(\mathbf{E}/\mathbf{K}_2)$, then for every $a \in \mathbf{K}_1$, $\phi^{-1}\psi\phi(a) = \phi^{-1}\psi[\phi(a)] = \phi^{-1}\phi(a) = a$, so that $\phi^{-1}\psi\phi \in G(\mathbf{E}/\mathbf{K}_1)$. Thus $G(\mathbf{E}/\mathbf{K}_1)$ and $G(\mathbf{E}/\mathbf{K}_2)$ are conjugate subgroups of $G(\mathbf{E}/\mathbf{F})$.

On the other hand, if $G(\mathbf{E}/\mathbf{K}_1) = \phi^{-1}H\phi$, where $H = G(\mathbf{E}/\mathbf{K}_2)$, we shall conclude that $\phi(\mathbf{K}_1) = \mathbf{K}_2$. For if $a \in \mathbf{K}_1$, but $\phi(a) \notin \mathbf{K}_2$, then $\phi(a)$ would not belong to the fixed field of H. Thus there would exist some $\psi \in H$ such that $\psi\phi(a) \neq \phi(a)$. But then $\phi^{-1}\psi\phi(a) \neq a$, contrary to the hypothesis. Thus $\phi(\mathbf{K}_1) \subset \mathbf{K}_2$. By a similar argument, $\phi^{-1}(\mathbf{K}_2) \subset \mathbf{K}_1$, and so $\mathbf{K}_2 \subset \phi(\mathbf{K}_1)$. Consequently, $\phi(\mathbf{K}_1) = \mathbf{K}_2$.

With respect to normality, recall that by the fundamental Galois pairing, there is a one-to-one correspondence between the intermediate fields and the subgroups of $G(\mathbf{E}/\mathbf{F})$. As shown above, conjugate fields correspond to conjugate subgroups. Now $\mathbf{K}$ is normal over $\mathbf{F}$ if and only if $\mathbf{K}$ has no conjugate field $\tilde{\mathbf{K}} \neq \mathbf{K}$. For if $\mathbf{K}$ is normal, it contains all the conjugates of any of its elements, by the definition of normality; and conversely, if $\mathbf{K}$ does contain all the conjugates of every one of its elements, it is the splitting field of a minimal polynomial of any r such that $\mathbf{K} = \mathbf{F}(r)$. Thus, $\mathbf{K}$ is normal over $\mathbf{F}$ if and only if $G(\mathbf{E}/\mathbf{K})$ has no conjugate subgroup other than itself; that is $\mathbf{K}$ is normal over $\mathbf{F}$ exactly when $G(\mathbf{E}/\mathbf{K})$ is a normal subgroup of $G(\mathbf{E}/\mathbf{F})$. ∎

PROBLEMS

1. If $\mathbf{F}_1 = \mathbf{F}(r_1)$ and $\mathbf{F}_2 = \mathbf{F}(r_2)$ and if $\mathbf{F}_1$ and $\mathbf{F}_2$ are conjugate fields over $\mathbf{F}$, does it follow that r_1 and r_2 are conjugates over $\mathbf{F}$?

2. If $\mathbf{E}$ is a finite extension of $\mathbf{F}$, show that there are only a finite number of intermediate fields $\mathbf{K}$, $\mathbf{E} \supset \mathbf{K} \supset \mathbf{F}$. (This already appeared as Problem 12 of Section 2.2, but an alternate and much simpler proof is now accessible.)

3. Show that conjugacy is an equivalence relation on the set of subgroups of a group. (Given a set S and a relation "$\sim$" holding between certain pairs of elements of S, this relation is called an *equivalence relation* if it is reflexive [for every $a \in S$, $a \sim a$], symmetric [if $a \sim b$, then also $b \sim a$] and transitive [if $a \sim b$ and $b \sim c$, then $a \sim c$].

4. Show that conjugacy is an equivalence relation on the set of finite extensions of a field $\mathbf{F}$.

5. If $\mathbf{E}$ is a normal extension of $\mathbf{F}$ and if $\mathbf{K}$ is an intermediate field, can every automorphism in $G(\mathbf{K}/\mathbf{F})$ be extended to an automorphism of $\mathbf{E}$?

6. Let **E** be an extension of **F**, and let ϕ be an automorphism of **F**. Can ϕ necessarily be extended to an automorphism of **E**: (a) in general? (b) if **E** is a finite extension of **F**? (c) if **E** is a normal extension of **F**?

7. Let r_1 and r_2 be conjugates over a field **F** and suppose that $r_2 \notin \mathbf{F}(r_1)$. Prove or disprove: $\mathbf{F}(r_1) \cap \mathbf{F}(r_2) = \mathbf{F}$.

Section 3.5. Galois' Theorem

At the beginning of this chapter, we undertook to study the question of when a polynomial is solvable by radicals. Let us review the major developments so far. First, the question was formulated in terms of fields. Letting f be a polynomial in x and letting **F** be the smallest field containing its coefficients, then the question is this: When is there a sequence of fields $\mathbf{F} = \mathbf{F}_0 \subset \mathbf{F}_1 \subset \cdots \subset \mathbf{F}_N$, each a radical extension of the previous one, such that the splitting field **E** of f satisfies $\mathbf{E} \subset \mathbf{F}_N$? Subsequent sections dealt with structural questions in the theory of field extensions and especially with their corresponding notions in the theory of groups. In particular, for any fields **E** and **F** such that **E** is a normal extension of **F**, a specific one-to-one correspondence was set up between intermediate fields **K**, $\mathbf{E} \supset \mathbf{K} \supset \mathbf{F}$, and subgroups of $G(\mathbf{E}/\mathbf{F})$. Under this correspondence, called the fundamental Galois pairing, those intermediate fields **K** which are themselves normal extensions of **F** were seen to correspond to subgroups $G(\mathbf{E}/\mathbf{K}) \subset G(\mathbf{E}/\mathbf{F})$ which are actually normal subgroups of $G(\mathbf{E}/\mathbf{F})$, and conversely. It has been asserted that this correspondence will enable us to treat problems about fields by studying the corresponding questions about groups, which are simpler. (One place where this has already been done is in Problem 2 of Section 3.4.) In this section we shall fully utilize this approach in order to answer the original question concerning solvability by radicals.

If G is a finite group, then G is said to be *solvable* if there is a sequence of subgroups $G = H_0 \supset H_1 \supset \cdots \supset H_N = \{\mathrm{id}\}$ such that each is a normal subgroup of prime index in the previous one; that is, if for each j, $0 \leqslant j \leqslant N-1$, H_{j+1} is a normal subgroup of H_j and $|H_j|/|H_{j+1}|$ is a prime number. Such a sequence of subgroups is called a *composition series for G with prime factors*, the prime factors being the numbers $|H_j|/|H_{j+1}|$. (Incidentally, if H_{j+1} is normal in H_j, this does not imply that it is normal in H_i for any i, $0 \leqslant i \leqslant j-1$. See Problem 9.) In this section we shall prove the famous **Theorem of Galois**:

THEOREM 34. *A polynomial is solvable by radicals if and only if its Galois group is solvable.*

Although the deepest results needed in the proof of this theorem have already been established in the previous section, there is still a fair amount of work to be done. Consequently, the proof presented here will be given in the form of a sequence of lemmas which should serve as an outline of the main ideas. The reader is advised not to become immersed in the details of the individual proofs before trying to understand the general form of the development.

Why should the solvability of the Galois group of a polynomial f have anything to do with the solvability of f by radicals? The best way to answer this seems to be to start on the "if" part of the theorem, treated in Lemmas 34a through 34d. Throughout, we shall denote by $\mathbf{F}$ the smallest field containing the coefficients of f and by $\mathbf{E}$ the splitting field of f.

LEMMA 34a. *If f has a solvable Galois group, then there is a sequence of fields $\mathbf{F} = \mathbf{F}_0 \subset \mathbf{F}_1 \subset \cdots \subset \mathbf{F}_N = \mathbf{E}$ such that for each j, $0 \leqslant j \leqslant N-1$, $\mathbf{F}_{j+1}$ is a normal extension of $\mathbf{F}_j$ of prime degree.*

Proof. By the solvability of the Galois group $G = G(\mathbf{E}/\mathbf{F})$, there is a sequence of subgroups $G = H_0 \supset H_1 \supset \cdots \supset H_N = \{\text{id}\}$, each a normal subgroup of prime index in the previous one. By the fundamental Galois pairing, Corollary 32, their corresponding fixed fields may be written $\mathbf{F} = \mathbf{F}_0 \subset \mathbf{F}_1 \subset \cdots \subset \mathbf{F}_N = \mathbf{E}$, where $\mathbf{F}_j$ is the fixed field of H_j. Being a splitting field of a polynomial over $\mathbf{F}$, $\mathbf{E}$ is a normal extension of $\mathbf{F}$, and so it is also a normal extension of each intermediate field $\mathbf{F}_j$ (Theorem 32); thus $H_j = G(\mathbf{E}/\mathbf{F}_j)$. Since H_{j+1} is a normal subgroup of H_j, Theorem 33 implies that $\mathbf{F}_{j+1}$ is a normal extension of $\mathbf{F}_j$. The degree of this extension is $[\mathbf{F}_{j+1} : \mathbf{F}_j] = [\mathbf{E} : \mathbf{F}_j]/[\mathbf{E} : \mathbf{F}_{j+1}] = |G(\mathbf{E}/\mathbf{F}_j)|/|G(\mathbf{E}/\mathbf{F}_{j+1})| = |H_j|/|H_{j+1}|$, which is the index of H_{j+1} in H_j and hence a prime. ∎

LEMMA 34b. *If $\tilde{\mathbf{K}}$ is a normal extension of prime degree p over a field $\mathbf{K}$ which contains the pth roots of unity, then there is a collection of $p-1$ numbers $a_1, a_2, \ldots, a_{p-1}$ such that for each i, $a_i^p \in \mathbf{K}$, and such that $\tilde{\mathbf{K}} \subset \mathbf{K}(a_1, a_2, \ldots, a_{p-1})$.*

Proof. The lemma simply asserts that every element in $\tilde{\mathbf{K}}$, a normal extension of prime degree over $\mathbf{K}$, can be obtained from elements of $\mathbf{K}$ using only rational operations and the extraction of pth roots.

Since $\tilde{\mathbf{K}}$ is normal over $\mathbf{K}$, we can write $\tilde{\mathbf{K}} = \mathbf{K}(r)$, where all the conjugates of r also belong to $\tilde{\mathbf{K}}$. We shall want to number these conjugates in a certain way. First observe that $G(\tilde{\mathbf{K}}/\mathbf{K})$ is of prime order p by normality; and it is cyclic, since by Lagrange's Theorem, the cyclic subgroup generated by any element $\phi \neq \text{id}$ must be the entire group. Thus we may write $G(\tilde{\mathbf{K}}/\mathbf{K}) = \{\phi, \phi^2, \ldots, \phi^p\}$. As noted many times before, ϕ must map conjugates of r to conjugates of r, because if g is the minimal polynomial of r over $\mathbf{K}$, then $0 = \phi(g(r)) = g(\phi(r))$. Define $r_i = \phi^i(r)$ for $1 \leqslant i \leqslant p$; then these r_i's must all be distinct because the ϕ^i's are all distinct and the action of ϕ^i on r uniquely determines its action on all of $\tilde{\mathbf{K}} = \mathbf{K}(r)$. Since $[\tilde{\mathbf{K}} : \mathbf{K}] = p$, r has exactly p conjugates, and so $r_1, r_2, \ldots, r_p$ are the entire set of roots of g. Letting ω be a primitive pth root of unity, define

$$a_j = r_1 + \omega^j r_2 + \omega^{2j} r_3 + \cdots + \omega^{(p-1)j} r_p$$

for $0 \leqslant j \leqslant p-1$. Writing these out for the different values of j, we obtain p linear equations in r_1 through r_p:

$$\begin{aligned}
a_0 &= r_1 + r_2 + r_3 + \cdots + r_p \\
a_1 &= r_1 + \omega r_2 + \omega^2 r_3 + \cdots + \omega^{(p-1)} r_p \\
a_2 &= r_1 + \omega^2 r_2 + \omega^4 r_3 + \cdots + \omega^{(p-1)2} r_p \\
&\vdots \\
a_{p-1} &= r_1 + \omega^{(p-1)} r_2 + \omega^{2(p-1)} r_3 + \cdots + \omega^{(p-1)^2} r_p
\end{aligned}$$

Considering the r_i's as the 'unknowns', we will show that this is a nonsingular system of equations and that for each j, $(a_j)^p \in \mathbf{K}$. Once the former fact has been established, it will follow from the elementary theory of linear equations that the r_i's may be obtained from the numbers a_j and ω via rational operations.

First observe that $a_0 \in \mathbf{K}$ since it is an (elementary) symmetric polynomial evaluated at the roots of a polynomial g over $\mathbf{K}$. For each j, $1 \leqslant j \leqslant p-1$, we show that $(a_j)^p \in \mathbf{K}$ by showing that $(a_j)^p$ is unchanged by every automorphism of $G(\tilde{\mathbf{K}}/\mathbf{K})$, from which it follows that it is in the fixed field of $G(\tilde{\mathbf{K}}/\mathbf{K})$, which is $\mathbf{K}$. First we calculate $\phi[(a_j)^p]$:

$$\begin{aligned}
\phi\left[(a_j)^p\right] &= \left[\phi(a_j)\right]^p \\
&= \left[\phi(r_1) + \omega^j\phi(r_2) + \omega^{2j}\phi(r_3) + \cdots + \omega^{(p-1)j}\phi(r_p)\right]^p \\
&= \left[r_2 + \omega^j r_3 + \omega^{2j} r_4 + \cdots + \omega^{(p-1)j} r_1\right]^p \\
&= \left[\omega^{-j} a_j\right]^p \\
&= (a_j)^p.
\end{aligned}$$

Consequently, we see that for all i, $\phi^i[(a_j)^p] = (a_j)^p$. Thus, $(a_j)^p \in \mathbf{K}$.

To show that the system of equations is nonsingular, observe that the determinant of the coefficient matrix is a Vandermonde determinant:

$$\begin{vmatrix}
1 & 1 & 1^2 & 1^3 & \cdots & 1^{p-1} \\
1 & \omega & \omega^2 & \omega^3 & \cdots & \omega^{p-1} \\
1 & \omega^2 & (\omega^2)^2 & (\omega^2)^3 & \cdots & (\omega^2)^{p-1} \\
\vdots & & & & & \\
1 & \omega^{p-1} & (\omega^{p-1})^2 & (\omega^{p-1})^3 & \cdots & (\omega^{p-1})^{p-1}
\end{vmatrix}
= \prod_{0 \leqslant i < j \leqslant p-1} (\omega^j - \omega^i),$$

which is nonzero since $\omega^j \neq \omega^i$ for $0 \leqslant i < j \leqslant p-1$.

Since the system is nonsingular, the values of the r_i's can be calculated from the a_j's and ω by means of Cramer's Rule, for example, and hence by rational operations. Thus, for each i, $r_i \in \mathbf{K}(a_1, a_2, \ldots, a_{p-1})$; and so $\tilde{\mathbf{K}} = \mathbf{K}(r) \subset \mathbf{K}(a_1, a_2, \ldots, a_{p-1})$. ∎

In order to apply Lemma 34b, we need to modify the sequence of fields in Lemma 34a so that they contain the relevant roots of unity. The key to this step is the following:

LEMMA 34c. *If $\tilde{\mathbf{K}}$ is a normal extension of prime degree p over a field $\mathbf{K}$ and if ω is a primitive pth root of unity, then $\tilde{\mathbf{K}}(\omega)$ is a normal extension of degree p over $\mathbf{K}(\omega)$.*

Proof. Consider the following diagram:

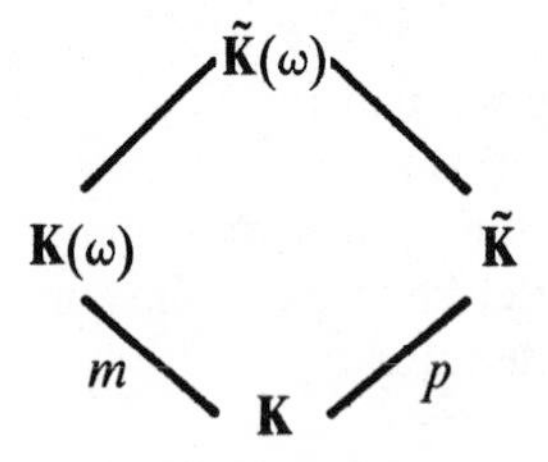

in which a line indicates that the upper field is an extension of the lower one, and where the letters m and p denote the degrees of those extensions. Since p divides $[\tilde{\mathbf{K}}(\omega) : \tilde{\mathbf{K}}][\tilde{\mathbf{K}} : \mathbf{K}] = [\tilde{\mathbf{K}}(\omega) : \mathbf{K}] = [\tilde{\mathbf{K}}(\omega) : \mathbf{K}(\omega)][\mathbf{K}(\omega) : \mathbf{K}]$, and since $[\mathbf{K}(\omega) : \mathbf{K}] = \deg_{\mathbf{K}}\omega \leqslant \deg_{\mathbf{Q}}\omega = p - 1$, it follows that p must divide $[\tilde{\mathbf{K}}(\omega) : \mathbf{K}(\omega)]$. Hence $[\tilde{\mathbf{K}}(\omega) : \mathbf{K}(\omega)] \geqslant p$. However $[\tilde{\mathbf{K}}(\omega) : \tilde{\mathbf{K}}] = \deg_{\tilde{\mathbf{K}}}\omega \leqslant \deg_{\mathbf{K}}\omega = m$, so that $p \geqslant [\tilde{\mathbf{K}}(\omega) : \mathbf{K}(\omega)]$. Thus $[\tilde{\mathbf{K}}(\omega) : \mathbf{K}(\omega)] = p$.

To see that $\tilde{\mathbf{K}}(\omega)$ is a normal extension of $\mathbf{K}(\omega)$, write $\tilde{\mathbf{K}} = \mathbf{K}(r)$ and suppose that g is the minimal polynomial of

r over $\mathbf{K}$. Then $\tilde{\mathbf{K}}(\omega)$ is the splitting field of $g(x)(x^p - 1)$ over $\mathbf{K}$, and hence it is normal over $\mathbf{K}$ and $\mathbf{K}(\omega)$ as well. (Alternatively, one could say that $\tilde{\mathbf{K}}(\omega)$ is the splitting field of g over $\mathbf{K}(\omega)$.) ∎

By putting together the preceding lemmas, we are able to complete the proof of the first half of the theorem:

LEMMA 34d. *If $G(\mathbf{E}/\mathbf{F})$ is solvable, then f is solvable by radicals.*

Proof. Consider an arbitrary pair $\mathbf{F}_j$ and $\mathbf{F}_{j+1}$ of successive fields in the sequence guaranteed by Lemma 34a. If $[\mathbf{F}_{j+1} : \mathbf{F}_j] = p$ and if ω is a primitive pth root of unity, then by Lemmas 34c and 34b, $\mathbf{F}_{j+1} \subset \mathbf{F}_j(\omega, a_1, a_2, \ldots, a_{p-1})$, where for each i, $a_i^p \in \mathbf{F}_j$. Using an additional subscript to keep track of which $\mathbf{F}_j$ we are working with, we could rewrite this as $\mathbf{F}_{j+1} \subset \mathbf{F}_j(\omega_j, a_{j,1}, a_{j,2}, \ldots, a_{j,p_j})$. Now, if we start with $\mathbf{F}$ and adjoin successively the numbers: $\omega_0, a_{0,1}, a_{0,2}, \ldots, a_{0,p_0}, \omega_1, a_{1,1}, a_{1,2}, \ldots, a_{1,p_1}, \ldots, \omega_{n-1}, \ldots, a_{n-1,p_{n-1}}$, we obtain a sequence of radical extensions such that the final one contains $\mathbf{F}_N$ and hence $\mathbf{E}$. Thus f is solvable by radicals. ∎

To prove the "only if" part of the theorem, we successively revise the sequence of radical extensions guaranteed by the definition of solvability by radicals until we obtain a sequence of normal extensions of prime degree such that the last one exactly equals $\mathbf{E}$, the splitting field. To begin, we have:

LEMMA 34e. *If f is solvable by radicals, there is a sequence of fields $\mathbf{F} = \mathbf{F}_0 \subset \mathbf{F}_1 \subset \cdots \subset \mathbf{F}_N$ such that*

$\mathbf{E} \subset \mathbf{F}_N$ and such that for each j, $\mathbf{F}_{j+1} = \mathbf{F}_j(a_j)$ where $a_j^{p_j} \in \mathbf{F}_j$ for some prime p_j.

Proof. In the original sequence guaranteed by the definition of solvability, take an arbitrary pair of successive fields $\mathbf{F}_j \subset \mathbf{F}_{j+1} = \mathbf{F}_j(a)$, where $a^n \in \mathbf{F}_j$. Let n be written as a product of primes, $n = p_1 p_2 \cdots p_k$. By adjoining to $\mathbf{F}_j$ the elements $a^{p_2 p_3 \cdots p_k}, a^{p_3 p_4 \cdots p_k}, \ldots, a^{p_k}$, a in succession, we obtain $\mathbf{F}_{j+1}$ through a sequence of radical extensions of the required type. If we insert such fields between every pair in the original sequence, we obtain a new sequence of the required type. (For convenience, let us agree to represent the new sequence by the same notation as the original.) ∎

Next we should like to modify this sequence of fields to a sequence of normal extensions of prime degree, but first there are some preliminaries.

LEMMA 34f. *If ω is a primitive pth root of unity, where p is a prime, and if $\mathbf{K}$ is any field, then $G(\mathbf{K}(\omega)/\mathbf{K})$ is cyclic.*

Proof. If $\phi \in G(\mathbf{K}(\omega)/\mathbf{K})$, $\phi(\omega)$ must be a conjugate of ω and hence must equal some power of ω. Moreover, ϕ is uniquely determined by its value at ω. For each $\phi \in G(\mathbf{K}(\omega)/\mathbf{K})$, let $\tilde{\phi}$ be the restriction of ϕ to the domain $\mathbf{Q}(\omega)$. Then $\tilde{\phi}$ is obviously an automorphism of $\mathbf{Q}(\omega)$ and hence an element of $G(\mathbf{Q}(\omega)/\mathbf{Q})$. Moreover, $\{\tilde{\phi} \mid \phi \in G(\mathbf{K}(\omega)/\mathbf{K})\}$ is a subgroup of the cyclic group $G(\mathbf{Q}(\omega)/\mathbf{Q})$, and hence it is itself cyclic. (See Problem 1, Section 3.3 and Problem 10, Section 3.2.) Let $\tilde{\psi}$ be a generator of this subgroup and let ψ be the corresponding element of $G(\mathbf{K}(\omega)/\mathbf{K})$, ψ being uniquely determined since it is determined by the value of $\tilde{\psi}(\omega)$. Then ψ is a generator of $G(\mathbf{K}(\omega)/\mathbf{K})$, since $\tilde{\phi} = (\tilde{\psi})^k$ implies that $\phi = (\psi)^k$, again because they are uniquely determined by their values at ω. ∎

Lemma 34f will be used in conjunction with Lemma 34g below in order to adjoin to **F**, by a sequence of normal extensions of prime degree, as many roots of unity as we shall want.

LEMMA 34g. *Every cyclic group is solvable.*

Proof. If G is a cyclic group of order n, we may write $G = \{\phi, \phi^2, \ldots, \phi^n = \text{id}\}$, where ϕ is a generator of G. If we write n as a product of primes, $n = p_1 p_2 \cdots p_k$, then the cyclic subgroups of G generated by $\phi, \phi^{p_1}, \phi^{p_1 p_2}, \ldots, \phi^{p_1 p_2 \cdots p_k}$, respectively, each have prime index in the previous one, for their orders are respectively $p_1 p_2 p_3 \cdots p_k, p_2 p_3 \cdots p_k, p_3 \cdots p_k, \ldots, 1$. Moreover, since G is Abelian (Theorem 30), each of its subgroups, and in particular these, must be normal, because $\phi^{-1} H \phi = \{\phi^{-1}\psi\phi \mid \psi \in H\} = \{\phi^{-1}\phi\psi \mid \psi \in H\} = \{\psi \mid \psi \in H\} = H$, for all $\phi \in G$. ∎

The last auxiliary result needed is the following:

LEMMA 34h. *If* **K** *contains the pth roots of unity, where p is a prime, and if* $a^p \in \mathbf{K}$ *but* $a \notin \mathbf{K}$, *then* $\mathbf{K}(a)$ *is a normal extension of degree p over* **K**.

Proof. If ω is a primitive pth root of unity, then the roots of $x^p - a^p$ are $a, \omega a, \omega^2 a, \ldots, \omega^{p-1} a$. Since these are all in $\mathbf{K}(a)$, $\mathbf{K}(a)$ is the splitting field over **K** of $x^p - a^p$; and so it is a normal extension of **K**. Since $a \notin \mathbf{K}$, $|G(\mathbf{K}(a)/\mathbf{K})| = [\mathbf{K}(a) : \mathbf{K}] \geqslant 2$, and so there exists an automorphism

$\phi \in G(\mathbf{K}(a)/\mathbf{K})$ such that $\phi(a) \neq a$. But $\phi(a)$ is a root of $x^p - a^p$, and so $\phi(a)/a$ is a pth root of unity; not being equal to one, this must be a primitive pth root of unity, say ω. It follows that $\phi^j(a) = \phi^{j-1}(\omega a) = \omega\phi^{j-1}(a) = \cdots = \omega^j a$ since $\phi(\omega) = \omega$ because $\omega \in \mathbf{K}$. Therefore the automorphisms $\phi, \phi^2, \phi^3, \ldots, \phi^p$ are all distinct since $\phi^j(a) = \omega^j a \neq \omega^i a = \phi^i(a)$ for $1 \leqslant i < j \leqslant p$. Consequently $|G(\mathbf{K}(a)/\mathbf{K})| \geqslant p$. At the same time $|G(\mathbf{K}(a)/\mathbf{K})| = [\mathbf{K}(a) : \mathbf{K}] = \deg_{\mathbf{K}} a \leqslant p$, so that $|G(\mathbf{K}(a)/\mathbf{K})| = p$. Therefore, $[\mathbf{K}(a) : \mathbf{K}] = p$. ∎

Now we are ready to move into the mainstream of the proof of the second half of Galois' Theorem.

LEMMA 34i. *If f is solvable by radicals, then there exists a sequence of fields $\mathbf{F} = \mathbf{E}_0 \subset \mathbf{E}_1 \subset \cdots \subset \mathbf{E}_M$ such that $\mathbf{E} \subset \mathbf{E}_M$ and such that for each j, $\mathbf{E}_{j+1}$ is a normal extension of prime degree over $\mathbf{E}_j$.*

Proof. Returning to the sequence of fields given by Lemma 34e, let us make a list of all the a's and p's occurring in that sequence: $p_0, a_0, p_1, a_1, p_2, a_2, \ldots, p_{N-1}, a_{N-1}$. Let ω_i denote a primitive p_ith root of unity. By starting at $\mathbf{F}$ and adjoining successively $\omega_0, a_0, \omega_1, a_1, \omega_2, a_2, \ldots, \omega_{N-1}, a_{N-1}$, we shall be able to obtain the desired sequence of field extensions. This is really very simple. Each time we adjoin an ω_i, the group of the extension is cyclic (Lemma 34f) and hence solvable (Lemma 34g). Since the extension is actually the splitting field of the p_ith cyclotomic polynomial, by Lemma 34a it can be obtained by a sequence of normal extensions of prime degree. Each time we wish to adjoin an a_i, since ω_i must already have been adjoined, either the extension is of degree one (i.e., a_i already belongs to the previous field), or else Lemma 34h guarantees that the extension is a single normal extension of prime degree p_i.

To formalize these ideas, we proceed by induction on N, the number of extensions given by Lemma 34e. If $N = 0$, we are already done; no extensions are necessary. As the inductive hypothesis, we now assume that for $N = R$ it is possible to modify the sequence $\mathbf{F}_0 \subset \mathbf{F}_1 \subset \cdots \subset \mathbf{F}_R$ into the desired type of sequence $\mathbf{E}_0 \subset \mathbf{E}_1 \subset \cdots \subset \mathbf{E}_M$ such that $\mathbf{F}_R \subset \mathbf{E}_M$. For $N = R + 1$, we simply want to adjoin ω_R and a_R to $\mathbf{E}_M$. If $\omega_R \in \mathbf{E}_M$, we go on to a_R. If $\omega_R \notin \mathbf{E}_M$, by Lemma 34f, $G(\mathbf{E}_M(\omega_R)/\mathbf{E}_M)$ is cyclic; by Lemma 34g it is solvable. Since $\mathbf{E}_M(\omega_R)$ is the splitting field over $\mathbf{E}_M$ of the p_Rth cyclotomic polynomial, $(x^{p_R} - 1)/(x - 1) = x^{p_R - 1} + x^{p_R - 2} + \cdots + x + 1$, Lemma 34a guarantees that there is a sequence $\mathbf{E}_M \subset \mathbf{E}_{M+1} \subset \cdots \subset \mathbf{E}_J = \mathbf{E}_M(\omega_R)$, in which each field is a normal extension of prime degree over the previous one. Next we adjoin a_R. If $a_R \in \mathbf{E}_J$, we are already done. If $a_R \notin \mathbf{E}_J$, by Lemma 34h, $\mathbf{E}_{J+1} = \mathbf{E}_J(a_R)$ is a normal extension of prime degree p_R over $\mathbf{E}_J$. Thus the sequence $\mathbf{E}_0 \subset \mathbf{E}_1 \subset \cdots \subset \mathbf{E}_M$ has been extended to include ω_R and a_R, as desired. ∎

Finally we wish to modify the sequence of fields given in Lemma 34i so that the last one actually equals $\mathbf{E}$, the splitting field of f.

LEMMA 34j. *If f is solvable by radicals, then there exists a sequence of fields $\mathbf{F} = \tilde{\mathbf{E}}_0 \subset \tilde{\mathbf{E}}_1 \subset \cdots \subset \tilde{\mathbf{E}}_J$ such that $\mathbf{E} = \tilde{\mathbf{E}}_J$ and such that for each j, $\tilde{\mathbf{E}}_{j+1}$ is a normal extension of prime degree over $\tilde{\mathbf{E}}_j$.*

Proof. Referring to the fields given in Lemma 34i, define $\mathbf{K}_j = \mathbf{E}_j \cap \mathbf{E}$. Thus we have $\mathbf{F} = \mathbf{K}_0 \subset \mathbf{K}_1 \subset \cdots$

$\subset \mathbf{K}_M = \mathbf{E}$, and it remains to show that once repetitions have been omitted from this sequence, it will be exactly what we want. Consider the following diagram:

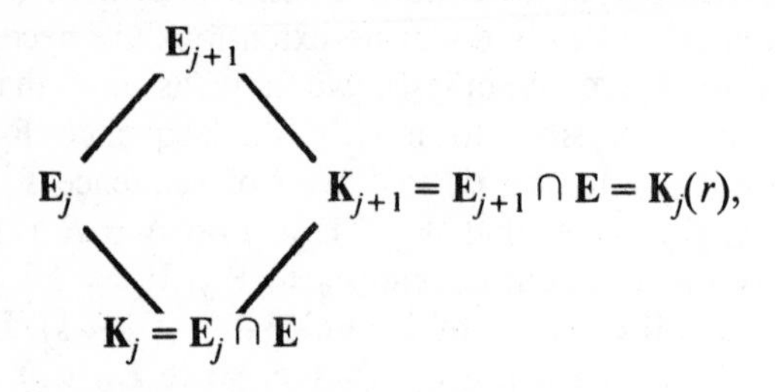

where a line indicates that the upper field is an extension of the lower one. We have written $\mathbf{K}_{j+1}$ as a simple algebraic extension $\mathbf{K}_j(r)$. First we observe that $\mathbf{K}_{j+1}$ is a normal extension of $\mathbf{K}_j$. For, suppose $g \in \mathbf{K}_j[x]$ has one root $s \in \mathbf{K}_{j+1}$; since $g \in \mathbf{E}_j[x]$ and $s \in \mathbf{E}_{j+1}$, g has all its roots in $\mathbf{E}_{j+1}$, by the normality of $\mathbf{E}_{j+1}$ over $\mathbf{E}_j$. Similarly, since $g \in \mathbf{E}[x]$, and since $s \in \mathbf{E}$, a normal extension of itself, all the roots of g must be in $\mathbf{E}$. Since the roots must be in both $\mathbf{E}_{j+1}$ and $\mathbf{E}$, they must be in $\mathbf{K}_{j+1}$. In particular, all the conjugates of r over $\mathbf{K}_j$ must be in $\mathbf{K}_{j+1}$.

Now, if $r \in \mathbf{E}_j$ then $r \in \mathbf{K}_j$, so that $\mathbf{K}_{j+1} = \mathbf{K}_j$. If $r \notin \mathbf{E}_j$, then $\mathbf{E}_{j+1} = \mathbf{E}_j(r)$; for $\mathbf{E}_j \subsetneqq \mathbf{E}_j(r) \subset \mathbf{E}_{j+1}$. Since $[\mathbf{E}_j(r) : \mathbf{E}_j]$ divides the prime $p = [\mathbf{E}_{j+1} : \mathbf{E}_j]$, $[\mathbf{E}_j(r) : \mathbf{E}_j] = p$. By normality of $\mathbf{E}_{j+1}$ over $\mathbf{E}_j$, there are exactly p automorphisms in $G(\mathbf{E}_{j+1}/\mathbf{E}_j)$, and each maps r to a distinct conjugate of r over $\mathbf{E}_j$. Since conjugates of r over $\mathbf{E}_j$ are certainly conjugates of r over $\mathbf{K}_j$, each $\tilde{\phi}$ obtained by restricting $\phi \in G(\mathbf{E}_{j+1}/\mathbf{E}_j)$ to $\mathbf{K}_{j+1}$ is an automorphism belonging to $G(\mathbf{K}_{j+1}/\mathbf{K}_j)$. (Cf. proof of Lemma 33, Problem 5, Section 3.4, and the proof of Lemma 34f.) Since each ϕ maps r to a different conjugate, there result in this way p distinct elements $\tilde{\phi} \in G(\mathbf{K}_{j+1}/\mathbf{K}_j)$. These elements clearly form a subgroup H of order p of $G(\mathbf{K}_{j+1}/\mathbf{K}_j)$, and it is our aim to show that there are no other elements of $G(\mathbf{K}_{j+1}/\mathbf{K}_j)$. By the one-to-one nature of the Galois pairing, it suffices to show that $\mathbf{K}_j$ is the fixed field of H. Suppose this were not the case. Then there would exist a number $a \in \mathbf{K}_{j+1}$, $a \notin \mathbf{K}_j$, such that $\tilde{\phi}(a) = a$ for all $\tilde{\phi} \in H$. Since $\mathbf{K}_j = \mathbf{E}_j \cap \mathbf{K}_{j+1}$, it must be that $a \notin \mathbf{E}_j$, the fixed field of $G(\mathbf{E}_{j+1}/\mathbf{E}_j)$. Hence there exists a $\phi \in G(\mathbf{E}_{j+1}/\mathbf{E}_j)$ such that $\phi(a) \neq a$; thus $\tilde{\phi}(a) = \phi(a) \neq a$, a contradiction.

Therefore H, of order p, is $G(\mathbf{K}_{j+1}/\mathbf{K}_j)$. Since $\mathbf{K}_{j+1}$ is a normal extension of $\mathbf{K}_j$, $[\mathbf{K}_{j+1} : \mathbf{K}_j] = p$.

We have shown then that there are only two possibilities at each step: either $\mathbf{K}_{j+1} = \mathbf{K}_j$ or $\mathbf{K}_{j+1}$ is a normal extension of prime degree over $\mathbf{K}_j$. By listing the **K**'s, but excluding the repetitions, we obtain the sequence of fields sought. ∎

From here the final step in the proof of Theorem 34 is almost immediate.

LEMMA 34k. *If f is solvable by radicals, then its Galois group is solvable.*

Proof. Corresponding to the fields given in Lemma 34j, the fundamental Galois pairing applied to the normal extension **E** over **F** produces the normal subgroups $G(\mathbf{E}/\mathbf{F})$, $G(\mathbf{E}/\tilde{\mathbf{E}}_1)$, $G(\mathbf{E}/\tilde{\mathbf{E}}_2), \ldots, G(\mathbf{E}/\tilde{\mathbf{E}}_J) = \{\mathrm{id}\}$. The index of $G(\mathbf{E}/\tilde{\mathbf{E}}_{j+1})$ in $G(\mathbf{E}/\tilde{\mathbf{E}}_j)$ is $|G(\mathbf{E}/\tilde{\mathbf{E}}_j)|/|G(\mathbf{E}/\tilde{\mathbf{E}}_{j+1})|$ $= [\mathbf{E} : \tilde{\mathbf{E}}_j]/[\mathbf{E} : \tilde{\mathbf{E}}_{j+1}] = [\tilde{\mathbf{E}}_{j+1} : \tilde{\mathbf{E}}_j]$, which is prime. Thus $G(\mathbf{E}/\mathbf{F})$, the Galois group of f, must be solvable! ∎

Now that we have a necessary and sufficient condition for solvability by radicals, it is natural to ask what can be done with it. After all, up to this point we have not even calculated the Galois groups of any but the most trivial polynomials; and if you can't calculate the Galois group of a polynomial, you could hardly expect to be able to tell whether it's solvable or not. The calculation of the Galois group of $x^4 - 2$ is developed in the problems. In the next section, we shall explicitly calculate the Galois group for a certain class of polynomials, and we shall show that these particular groups are not solvable. From this, of course, will follow the fact that there exist polynomials which are not solvable by radicals. In the subsequent section, we shall treat several polynomials whose groups are solvable; and by reconstructing the ideas in the proof of Theorem 34, we shall actually show how to solve them by radicals.

PROBLEMS

1. If $\mathbf{E}$ is a normal extension of $\mathbf{F}$ and if $\mathbf{K}_1$ and $\mathbf{K}_2$ are intermediate fields, describe $G(\mathbf{E}/\mathbf{K}_1 \cap \mathbf{K}_2)$.
2. In the context of Problem 1, show that if $\mathbf{K}_1$ and $\mathbf{K}_2$ are actually normal extensions of $\mathbf{F}$, then $G(\mathbf{E}/\mathbf{K}_1 \cap \mathbf{K}_2) = \{\phi\psi \mid \phi \in G(\mathbf{E}/\mathbf{K}_1), \psi \in G(\mathbf{E}/\mathbf{K}_2)\}$.
3. If $\mathbf{E}$ is a normal extension of $\mathbf{F}$ with intermediate fields $\mathbf{K}_1$ and $\mathbf{K}_2$, and if $\mathbf{K}$ is the smallest field containing $\mathbf{K}_1$ and $\mathbf{K}_2$, what is $G(\mathbf{E}/\mathbf{K})$?
4. Let ω be a primitive fifth root of unity. Find a field $\mathbf{F}$ such that $[\mathbf{F}(\omega) : \mathbf{F}] = 2$.
5. Let ω be a primitive 31st root of unity. Show that there exists a field $\mathbf{F}$ such that $[\mathbf{F}(\omega) : \mathbf{F}] = 6$.
6. Find the order of the Galois group of $x^4 - 2$; and then describe completely, in some explicit way, each of its elements.

7. Find all the subgroups of the group obtained in the previous problem, and determine which ones are normal subgroups.

8. If **E** is a normal extension of **K** and if **K** is a normal extension of **F**, does it follow that **E** is a normal extension of **F**?

9. If H is a normal subgroup of a group G and if N is a normal subgroup of H, does it follow that N is a normal subgroup of G?

10. Why should it be obvious that the Galois group calculated in Problem 6 is solvable? Verify its solvability directly.

11. If **F** is a field containing no pth roots of the number $A \in \mathbf{F}$, for a fixed prime p, show that the polynomial $x^p - A$ is irreducible over **F**.

Section 3.6. Abel's Theorem

In this section we shall prove that there exist polynomials which are not solvable by radicals. This will be accomplished by explicitly calculating the Galois groups of a whole class of polynomials and then by showing that the resulting groups are not solvable. As was noted at the end of the previous section, we have not spent much effort up to now on the question of how to find, or even how to describe effectively, the Galois group of a polynomial. For certain very simple cases such groups were described in the text; in Problem 1 of Section 3.3 the Galois group of the pth cyclotomic polynomial was calculated; and in Problem 6 of Section 3.5 the Galois group of $x^4 - 2$ was calculated. But in all these cases considerable knowledge about the roots of the polynomials was used in describing their groups. In this section we shall see an example of how it is possible to describe and calculate Galois groups without much prior knowledge of the roots of the relevant polynomials.

Let us assume that f is a polynomial of degree n over a field $\mathbf{F}$ and that the roots of f are labelled in an arbitrary but fixed way as $r_1, r_2, \ldots, r_n$. Let us further assume that these roots are all distinct. (This will be true, for example, in the special case when f is irreducible over $\mathbf{F}$.) Letting $\mathbf{E}$ be the splitting field over $\mathbf{F}$ of f, we have $\mathbf{E} = \mathbf{F}(r_1, r_2, \ldots, r_n)$. Recall that the Galois group $G = G(\mathbf{E}/\mathbf{F})$ is the group of all automorphisms of $\mathbf{E}$ which leave each element of $\mathbf{F}$ unchanged. Because of the fact that if $\phi \in G, f(r) = 0$ implies that $\phi[f(r)] = f[\phi(r)] = 0$, each automorphism of G maps the set of roots of f onto itself; that is, each $\phi \in G$ simply induces a permutation on the numbers $r_1, r_2, \ldots, r_n$. Moreover, the action of ϕ on these roots completely determines its action on all of $\mathbf{E}$, for any element of $\mathbf{E}$ can be expressed as a polynomial combination of these roots. Therefore, any $\phi \in G$ can be described completely by describing the permutation which it induces on the r_i's.

There happens to be a convenient standard way to represent permutations. To begin, suppose $\phi \in G$ and let r_j be one of the roots of f. We can start to make a list: $r_j, \phi(r_j), \phi^2(r_j) \ldots$. Eventually some $\phi^k(r_j)$ must repeat something earlier in the list, since there are only n possible values for the numbers in the list to assume. If k corresponds to the exponent of ϕ in the first such repetition, then $\phi^k(r_j) = r_j$ itself; for if $\phi^k(r_j) = \phi^i(r_j)$ with $i \geqslant 1$, then $\phi^{k-i}(r_j) = r_j$ would correspond to an earlier repetition. The list of numbers $r_j, \phi(r_j), \phi^2(r_j), \ldots, \phi^{k-1}(r_j)$ is called a *cycle*, for each entry is obtained by applying ϕ to the one before it, and the first is gotten by applying ϕ to the last. Now, if r_m is another root of f not in the cycle just constructed, we can construct in the same way a cycle containing r_m; and obviously the set of roots occurring in this cycle will be disjoint from the set of those occurring in the previous one. Continuing in this fashion, the action of ϕ on the roots of f can be described by a list of disjoint cycles. In writing these cycles, let us agree to represent the root r_j simply by writing the number j; further we shall enclose cycles in parentheses and juxtapose disjoint cycles.

For example, if f is an irreducible polynomial of degree 5 with (distinct) roots r_1, r_2, r_3, r_4, r_5, suppose that there is some $\phi \in G$, such that $\phi(r_1) = r_4$, $\phi(r_2) = r_1$, $\phi(r_3) = r_5$, $\phi(r_4) = r_2$, and $\phi(r_5) = r_3$. This information completely specifies ϕ. An abbreviation for ϕ would be (1 4 2)(3 5); for, reading from the left, it says that r_1 is mapped to r_4, r_4 to r_2, r_2 to r_1; also r_3 is mapped to r_5 and r_5 to r_3. There is some leeway in this system of abbreviation; in particular, *disjoint* cycles may be juxtaposed in either order, and any cycle may be begun with any of its elements. Thus, another representation of the ϕ given above would be (5 3)(2 1 4). As a further abbreviation, in writing the cycles representing a given automorphism, we shall agree to omit any cycles consisting of a single element; such cycles correspond to roots which are left fixed by the automorphism. For example, the automorphism represented by (5 3 2) maps r_5 to r_3, r_3 to r_2, r_2 to r_5 and it leaves r_1 and r_4 each unchanged. A cycle containing exactly k elements is called a *k-cycle*. A 2-cycle is called a *transposition*.

With the obvious meaning, we shall use the equality sign between the name of an automorphism and its representation. Thus, we could write $\phi = (5\ 3\ 2)$ to denote the automorphism described by this 3-cycle. With this kind of identification in mind, we shall say that a group *contains a k-cycle* if there is an element of the group *equal to* some k-cycle (as opposed to an element of the group simply containing a k-cycle).

The notation just introduced is convenient for computing the product of automorphisms. For example, in the case $n=5$, suppose $\phi=(1\ 3\ 4\ 2)$ and $\psi=(2\ 5)(4\ 3\ 1)$; we can compute

$$\begin{aligned}\phi\psi &= [(1\,3\,4\,2)][(2\,5)(4\,3\,1)]\\ &= (1\,2\,5)\end{aligned}$$

by tracing the effect on each number through the relevant cycles as these are encountered from right to left. That is, one might proceed as follows: Pick any number, say 1; by the first cycle on the right it is mapped to 4; moving now to the left until the next occurrence of 4, we see that 4 is mapped to 2. Thus 1 is ultimately mapped to 2 by the product. Next, we need the effect of the product on 2; it is mapped to 5 by $(2\ 5)$ and then 5 is left fixed; thus 2 is mapped to 5. Continuing in this way, we find that 5 is mapped to 1 and both 3 and 4 are left fixed. As another example, the reader should verify that

$$\begin{aligned}\psi\phi &= [(2\,5)(4\,3\,1)][(1\,3\,4\,2)]\\ &= (2\,4\,5).\end{aligned}$$

It is easy to see that the inverse of any automorphism is simply obtained by reversing the order within each of its cycles. Thus, if $\psi=(2\ 5)(4\ 3\ 1)$, then $\psi^{-1}=(5\ 2)(1\ 3\ 4)$.

In the case of a polynomial f of degree n with all its roots distinct, we have seen that to each element of the Galois group there corresponds a permutation on the n roots, or equivalently, a permutation on the subscripts 1 through n. It is customary to denote by S_n the set of all $n!$ permutations on the numbers 1 through n. If every such permutation occurs under some element of the group G, we say that $G = S_n$ or that G is the *symmetric group* on n objects. In many cases, however, not all permutations occur under elements of G. For example, we know (Problem 3 of Section 3.3) that the Galois group of $x^5 - 1$ has only four elements, even though there are $5! = 120$ possible permutations of the five roots.

Under certain circumstances, the Galois group of a polynomial will include all permutations of its roots. In particular, we have:

LEMMA 35a. *If f is an irreducible polynomial over* $\mathbf{Q}$, *of prime degree p, and if f has exactly $p - 2$ real roots, then the Galois group of f is S_p.*

Proof. Let the roots of f be $r_1, r_2, \ldots, r_p$, where r_1 and r_2 are the two complex, nonreal roots. The roots are all distinct since f is irreducible. First we shall show that the Galois group G is *transitive*, that is, for every pair of roots r_i and r_j, there exists some $\phi \in G$ such that $\phi(r_i) = r_j$. To see this, let $r_{i_1}, r_{i_2}, \ldots, r_{i_m}$ denote the complete set of distinct images of r_i under elements of G. Let P be the monic polynomial of degree m with roots $r_{i_1}, r_{i_2}, \ldots, r_{i_m}$. That is, $P(x) = \prod_{k=1}^{m}(x - r_{i_k})$. The coefficients of P are simply plus or minus the elementary symmetric functions of m variables, evaluated at these r's. We would like to show that this is actually a polynomial over $\mathbf{Q}$. To do this, note that the jth elementary symmetric function $\sigma_j(r_{i_1}, r_{i_2}, \ldots, r_{i_m})$ is unchanged by every $\phi \in G$, for

$$\phi[\sigma_j(r_{i_1}, r_{i_2}, \ldots, r_{i_m})] = \sigma_j(\phi(r_{i_1}), \phi(r_{i_2}), \ldots, \phi(r_{i_m}))$$

and the arguments on the right are distinct (since ϕ is one-to-one) and each is an image of r_i under an element of G. Since there are m of them, they must simply be a permutation of the list $r_{i_1}, r_{i_2}, \ldots, r_{i_m}$. Therefore, since σ_j is symmetric, its value is unchanged. Consequently, for each j, $\sigma_j(r_{i_1}, r_{i_2}, \ldots, r_{i_m}) \in \mathbf{Q}$, and so the numbers $r_{i_1}, r_{i_2}, \ldots, r_{i_m}$ are the roots of a polynomial of degree m over $\mathbf{Q}$. However, this polynomial obviously divides f in $\mathbf{Q}[x]$; so since f is irreducible, $m = p$. Thus the images of r_i exhaust the set of roots, and hence G is transitive. (So far we have only used the hypothesis that f is irreducible.)

Next we show that G contains the transposition (1 2). Since the complex roots of a polynomial with real coefficients occur in complex conjugate pairs, r_1 and r_2 must be complex conjugates. The automorphism of $\mathbf{C}$ defined by complex conjugation, when restricted to $\mathbf{E}$, must interchange r_1 and r_2 and leave the other roots, which are real, unchanged. Since this restriction is easily seen to map $\mathbf{E}$ onto $\mathbf{E}$, it is an element of G. Its representation is (1 2).

Now we show that G contains *every* transposition. To do this, define an equivalence relation on the set $S = \{1, 2, \ldots, p\}$ by $i \sim j$ if and only if $i = j$ or the transposition $(i\ j)$ belongs to G. This relation is obviously reflexive and symmetric; transitivity follows from the fact that if $(i\ j)$ and $(j\ k)$ belong to G, then $(j\ k)(i\ j)(j\ k) = (i\ k) \in G$. Let us denote by a bar the equivalence class of the element below it; a one-to-one correspondence between any two equivalence classes $\bar{i}$ and $\bar{j}$ can be obtained from any automorphism ϕ that maps r_i to r_j. To see this, let k represent an arbitrary element of S, $k \neq i$, and define m by the equation $\phi(r_k) = r_m$. Then $k \in \bar{i}$ if and only if $(i\ k) \in G$ if and only if $\phi(i\ k)\phi^{-1} = (j\ m) \in G$ if and only if $m \in \bar{j}$. So under this correspondence, i is associated with j and each $k \neq i$ is associated with the m

as defined above. Our equivalence relation divides S into disjoint equivalence classes, and now we know that each has the same number of elements. Since p is prime, either there are p equivalence classes of one element each or there is one equivalence class of p elements, namely S itself. The former case cannot hold, since both 1 and $2 \in \bar{1}$. Therefore G contains all transpositions. Since every permutation can be obtained by a succession of transpositions (Problem 7, Section 1.6), G must equal all of S_p. ∎

At the same time, we have:

LEMMA 35b. *If $n \geqslant 5$ and the Galois group $G = S_n$, then G is not solvable.*

Proof. First we establish the preliminary claim that if N is a normal subgroup of prime index p in a group H, then for any two elements $\phi, \psi \in H$, the product $\phi^{-1}\psi^{-1}\phi\psi \in N$. (The product $\phi^{-1}\psi^{-1}\phi\psi$ is called the *commutator* of ϕ and ψ.) If $\phi \in N$, by normality $\psi^{-1}\phi\psi \in N$, and then by closure $\phi^{-1}\psi^{-1}\phi\psi \in N$ also. If $\phi \notin N$, consider the set $\tilde{N} = \{\phi^k n \mid n \in N, k \in \mathbf{Z}\}$. It is a subgroup of H, since $\text{id} = \phi^0\text{id} \in \tilde{N}$; and $(\phi^k n)^{-1} = n^{-1}\phi^{-k} = \phi^{-k}\tilde{n} \in \tilde{N}$, for some $\tilde{n} \in N$ by the normality of N; and $(\phi^k n_1)(\phi^j n_2) = \phi^k\phi^j\tilde{n}_1 n_2 \in \tilde{N}$, again making use of normality. By Lagrange's Theorem $|\tilde{N}| = q|N|$ for some integer q, since N is a subgroup of $\tilde{N}$. At the same time, $|\tilde{N}|$ divides $|H| = p|N|$, and so $q = 1$ or $q = p$. Since $\phi \notin N$, $\tilde{N} \neq N$, and so $q \neq 1$. Thus $q = p$ and $\tilde{N} = H$. Consequently we may write $\psi = \phi^k n$ for some $k \in \mathbf{Z}$ and some $n \in N$. But then, $\phi^{-1}\psi^{-1}\phi\psi = \phi^{-1}n^{-1}\phi^{-k}\phi\phi^k n = \phi^{-1}n^{-1}\phi n \in N$, since $\phi^{-1}n^{-1}\phi \in N$ by normality and since N is closed.

Now suppose that G is solvable, so that there exists a sequence of subgroups $G = H_0 \supset H_1 \supset H_2 \supset \cdots \supset H_N = \{\text{id}\}$, where each H_{j+1} is a normal subgroup of prime index in H_j. We show that this is impossible, by showing that for every j, H_j contains every 3-cycle. (This yields a contradiction for $j = N$.) The proof is by induction. Since $G = H_0 = S_n$, H_0 contains every 3-cycle. Assuming now that H_j contains every 3-cycle, we show that H_{j+1} does also. Let $(i\,j\,k)$ be an arbitrary 3-cycle and define $\phi = (m\,j\,i)$ and $\psi = (i\,l\,k)$, where i, j, k, l, and m are all different (which is possible since $n \geqslant 5$); by the inductive hypothesis ϕ and $\psi \in H_j$. By the earlier claim, $\phi^{-1}\psi^{-1}\phi\psi \in H_{j+1}$. But if we compute $\phi^{-1}\psi^{-1}\phi\psi = (i\,j\,m)(k\,l\,i)(m\,j\,i)(i\,l\,k) = (i\,j\,k)$, we see that the arbitrary 3-cycle $(i\,j\,k)$ is in H_{j+1}. ∎

We may now prove our main result, **Abel's Theorem**:

THEOREM 35. *There exist polynomials of every degree* $\geqslant 5$ *which are not solvable by radicals.*

Proof. In the light of Lemmas 35a and 35b, we first seek an irreducible polynomial of degree 5 over **Q** which has exactly three real roots. The polynomial $f(x) = 2x^5 - 5x^4 + 5$, which was discussed in Section 3.1, is one such polynomial. Multiplying it by x^k, we obtain a polynomial of degree $5 + k$ which is also not solvable by radicals (because not *all* of its roots can be solved for in terms of radicals). ∎

In the case of the polynomial $2x^5 - 5x^4 + 5$, it is actually easy to see that *none* of its roots can be expressed in radicals. For if a single root r were to belong to a field **K** obtainable from **Q** by a sequence of radical extensions, then division of $2x^5 - 5x^4 + 5$ by $x - r$ would give a quartic over $\mathbf{K}$, and we have already seen (Problem 4, Section 3.1) that this must be solvable by radicals. Alternatively, the same result follows from the fact that in the case of an irreducible polynomial, either none or all of its roots may be expressed in terms of radicals (Problem 5).

PROBLEMS

1. Let ω be a primitive fifth root of unity, and label the roots of $x^5 - 1$ as: $r_1 = 1, r_2 = \omega, r_3 = \omega^2, r_4 = \omega^3, r_5 = \omega^4$. Give an example of a permutation of these roots which does not correspond to an element of the Galois group of the polynomial.
2. Label the roots of $x^4 - 2$ as: $r_1 = +\sqrt[4]{2}$ (positive real), $r_2 = i\sqrt[4]{2}$, $r_3 = -\sqrt[4]{2}$, $r_4 = -i\sqrt[4]{2}$. Give an example of a permutation of these roots which does not correspond to an element of the Galois group of the polynomial.
3. Prove or disprove: If $f \in \mathbf{F}[x]$ is irreducible, then its Galois group over $\mathbf{F}$ is transitive.
4. Find a fifth degree polynomial of the form $x^5 + bx + c$, such that it is not solvable by radicals.
5. If $f \in \mathbf{F}[x]$ is irreducible and there is a root r of f and a sequence of radical extensions $\mathbf{F} = \mathbf{F}_0 \subset \mathbf{F}_1 \subset \cdots \subset \mathbf{F}_N$ such that $r \in \mathbf{F}_N$, show that f is solvable by radicals over $\mathbf{F}$. (That is, if one root of an irreducible polynomial can be obtained by radicals, all the roots can be.)
6. Suppose that the group of a polynomial f over a field $\mathbf{F}$ is S_n. Show that f is irreducible over $\mathbf{F}$. Use this result to show that if in addition $n \geqslant 5$, then no root of f may be expressed in radicals.

7. Show that for every prime p, there exists a polynomial of degree p over $\mathbf{Q}$ which has S_p for its Galois group.

8. For every n, show that there exists a field $\mathbf{F}$ and an nth degree polynomial $f \in \mathbf{F}[x]$ such that the group of f over $\mathbf{F}$ is S_n. (Hint: Review Problems 6 and 7 of Section 3.3.)

Section 3.7. Some Solvable Equations

In the previous section we applied one of the two implications contained in Galois' Theorem in order to prove Abel's Theorem. In particular, we produced a polynomial with an unsolvable group, and therefore a polynomial not solvable by radicals. In this section, we shall apply the converse implication of Galois' Theorem, namely, that a polynomial with a solvable group is solvable by radicals. It should be mentioned at the outset that there actually do exist (rather complicated) procedures for calculating the Galois groups of large classes of polynomials, but we shall not pursue this general topic here. Instead, let us simply apply Galois' Theorem to several situations in which the group is known.

First, we observe that if ω is a primitive pth root of unity, p a prime, it can actually be calculated by a sequence of rational operations and the extraction of roots of lower order than p. To see this, recall that the Galois group G of the pth cyclotomic polynomial $f(x) = x^{p-1} + x^{p-2} + \cdots + 1$ is cyclic and of order $p - 1$ (Problem 1 of Section 3.3). Consequently, G is solvable and has a composition series whose factors p_i are the prime factors of $p - 1$. (See Lemma 34g and its proof.) As in the proof of Lemma 34d, f can be solved by means of radicals of order p_i, as p_i ranges over all the composition factors.

From the observations of the previous paragraph, we immediately obtain a short proof of Theorem 28, which asserts that the regular n-gon is constructible if the odd factors of n are distinct Fermat primes. For, as noted at the outset of the original proof, it suffices to consider the case where $n = p$, a Fermat prime; and so we want to show that a primitive pth root of unity is constructible. Well, since $p - 1$ is a power of 2, the pth cyclotomic polynomial may be solved using radicals only of order 2. Since these operations correspond to admissible constructions, the pth roots of unity are constructible in this case!

It is interesting to relate the original proof of Theorem 28 to the one just given. For convenience, let us treat the case $p = 17$, although the ideas are completely general. The Galois group of the 17th cyclotomic polynomial is cyclic and of order 16, and so it may be represented as $G = \{\phi, \phi^2, \ldots, \phi^{16} = \mathrm{id}\}$, where ϕ is determined by the condition $\phi(\omega) = \omega^g$ in which g is a primitive root modulo 17 and ω is a primitive 17th root of unity. Thus a complete list of the roots of the 17th cyclotomic polynomial is:

$$\phi(\omega), \phi^2(\omega), \phi^3(\omega), \ldots, \phi^{16}(\omega).$$

A composition series for G is $G = H_0 \supset H_1 \supset H_2 \supset H_3 \supset H_4 = \{\phi^{16} = \mathrm{id}\}$, where $H_1 = \{\phi^2, \phi^4, \phi^6, \phi^8, \phi^{10}, \phi^{12}, \phi^{14}, \phi^{16}\}$, $H_2 = \{\phi^4, \phi^8, \phi^{12}, \phi^{16}\}$, and $H_3 = \{\phi^8, \phi^{16}\}$. Under the fundamental Galois pairing, there exists a sequence of fields $\mathbf{Q} = \mathbf{F}_0 \subset \mathbf{F}_1 \subset \mathbf{F}_2 \subset \mathbf{F}_3 \subset \mathbf{F}_4 = \mathbf{Q}(\omega)$, such that each is a normal extension of degree 2 over the previous one (that is, a quadratic extension) and such that each $\mathbf{F}_i$ is the fixed field of H_i. Let us actually calculate the fields. We shall see that each $\mathbf{F}_j$ is the extension of $\mathbf{F}_{j-1}$ by one (and, in fact, all) of the j-periods, and that complementary j-periods are conjugates over $\mathbf{F}_{j-1}$.

For example, consider the 1-periods, which we can write here as

$$\eta_1 = \phi(\omega) + \phi^3(\omega) + \cdots + \phi^{15}(\omega),$$

$$\eta_2 = \phi^2(\omega) + \phi^4(\omega) + \cdots + \phi^{16}(\omega).$$

Since they are both left unchanged by the elements of H_1, they both belong to $\mathbf{F}_1$. By the linear independence over $\mathbf{Q}$ of the numbers $\omega^1 \ldots \omega^{16}$, which form a basis for $\mathbf{Q}(\omega)$ over $\mathbf{Q}$, $\eta_1 \neq \eta_2$. As a consequence, they are not left fixed by some elements of H_0, and so they are not in $\mathbf{F}_0$. In fact, since η_1 and η_2 are either left unchanged or mapped to each other by the elements of H_0, they are a complete set of conjugates over $\mathbf{F}_0$. Therefore they are the roots of a quadratic polynomial over $\mathbf{F}_0$. By exactly the same reasoning, the reader should be able to carry the argument through to subsequent extensions.

PROBLEMS

1. What is the group of an irreducible quadratic? Use the definition of a solvable group to show that it is a solvable group. Then carry through the procedure used in the first part of the proof of Galois' Theorem to derive an actual formula for its solution, a formula that should turn out to be the quadratic formula.

2. Repeat the previous problem for a cubic equation under the assumption that the group is S_3. Does your solution depend on the fact that the group is all of S_3? (Cf. Problem 1 of Section 3.1.)

3. Let f be a cubic polynomial with real coefficients and let $\mathbf{F}$ be the smallest field containing the coefficients. If r_1, r_2, and r_3 denote the roots of f, we define the

discriminant of f to be $\Delta^2 = \prod_{i<j}(r_i - r_j)^2$. Show that $\Delta^2 \in \mathbf{F}$. Show further that if $\Delta^2 > 0$ all the roots are real and if $\Delta^2 < 0$ there is exactly one real root. What if $\Delta^2 = 0$?

4. If r is any root of an irreducible cubic f, over some field $\mathbf{F}$, and if Δ is a square root of the discriminant, show that the splitting field $\mathbf{E}$ of f, over $\mathbf{F}$, is given by $\mathbf{F}(\Delta, r)$.

*5. Let f be an irreducible cubic over $\mathbf{Q}$ with three real roots. Show that it is not possible to solve for any of its roots by real radicals alone. (This may be surprising.)

6. Use Galois' Theorem to show that every quartic is solvable by radicals. (Cf. Problem 4, Section 3.1.)

REFERENCES AND NOTES

Early approaches to low degree polynomial equations are discussed in [1, 2]. Edited translations of some of the original writings are also available [3, 4]. The papers of Abel and Galois are available in these authors' collected works [5, 6], and parts have been translated into English [4]. The work of Abel and of Ruffini on the quartic equation is surveyed in [1] and discussed in depth in [7]. Older books on the theory of equations, such as [8, 9, 10], are largely based on the point of view of the original workers, which emphasized the notion that the group of an equation may be considered to be a set of permutations of its roots. Our treatment considered the group as a set of automorphisms, and it was only relatively late that we identified this concept with a set of permutations. This modern viewpoint is largely due to the influence of Artin, as in [11], but it has its foundation as far back as Dedekind (cf. [12]). The book [1] and the long paper [13] are very helpful in tracing the evolution of Galois theory into its modern form.

It has been mentioned from time to time that more general definitions of groups and fields may be given. For example, there are fields which are not sets of complex numbers and groups which make no reference to automorphisms of fields. Abstract algebra is the name given to treatments of such topics, and there are many excellent textbooks available. Some of these are [14–20]. While Galois theory is often developed in this more general setting, the reader should be cautioned that some of the results in this book do not carry over verbatim. For example, over more general fields, irreducible polynomials do not always have distinct roots. Books devoted specifically to Galois theory include [21, 22, 23]. For information on how to construct the group of an equation whose roots are not known, one might consult [23] or the early books on the theory of equations.

Although it may be complicated to analyze the roots of certain equations by the usual algebraic operations and the extraction of roots, perhaps they might be solved easily by other methods. For example, the representation of the roots of $x^n - 1$ by trigonometric functions was seen to be extremely simple. This idea was applied to the quintic equation by Hermite and Kronecker, who solved it in terms of elliptic functions; see p. 763 of [1].

The Artin quote in Section 3.2 is from p. 54 of [24].

1. Morris Kline, *Mathematical Thought from Ancient to Modern Times*, Oxford University Press, New York, 1972.

2. David Eugene Smith, *History of Mathematics*, vols. 1 and 2, Dover, New York, 1958.

3. Girolamo Cardano, *The Great Art or the Rules of Algebra*, M.I.T. Press, Cambridge, Mass., 1968 (orig. ed. 1545).

4. David Eugene Smith, ed., *A Source Book in Mathematics*, vols. 1 and 2, Dover, New York, 1959.

5. Niels Henrik Abel, *Oeuvres Complètes*, 2 vols., ed L. Sylow and S. Lie, Grondahl and Sons, Christiania, 1881.

6. Évariste Galois, *Ecrits et Mémoires Mathématiques*, ed. R. Bourgne and J.-P. Azra, Gauthier-Villars, Paris, 1962.

7. James Pierpont, On the Ruffini-Abelian Theorem, *Bull. Amer. Math. Soc.*, 2 (1895) 200–221.

8. Florian Cajori, *An Introduction to the Theory of Equations*, Dover, New York, 1969 (orig. ed. 1904).

9. Edgar Dehn, *Algebraic Equations*, Columbia University Press, New York, 1930.

10. L. E. Dickson, *Introduction to the Theory of Equations*, Wiley, New York, 1903.

11. Emil Artin, *Galois Theory*, Notre Dame Mathematical Lectures, Notre Dame, Ind., 1942.

12. Richard Dedekind, Über die Permutationen der Körpers aller algebraischen Zahlen (1901), *Gesammelte Mathematische Werke*, Chelsea, New York, 1969.

13. B. Melvin Kiernan, The development of Galois Theory from Lagrange to Artin, *Arch. History Exact Sci.*, 8 (1971–72) 40–154.

14. Garrett Birkhoff and Saunders Mac Lane, *A Survey of Modern Algebra*, 4th ed., Macmillan, New York, 1977.

15. Allan Clark, *Elements of Abstract Algebra*, Wadsworth, Belmont, Calif., 1971.

16. Larry Joel Goldstein, *Abstract Algebra: A First Course*, Prentice-Hall, Englewood Cliffs, N.J., 1973.

17. I. N. Herstein, *Topics in Algebra*, 2nd ed., Xerox, Lexington, Mass., 1975.

18. Abraham P. Hillman and Gerald L. Alexanderson, *A First Undergraduate Course in Abstract Algebra*, Wadsworth, Belmont, Calif., 1973.

19. B. L. van der Waerden, *Modern Algebra*, vol. 1, Frederick Ungar, New York, 1949 (orig. ed. 1931).

20. Helmut Hasse, *Higher Algebra*, vol. 1 and 2, tr. by T. Benac, Frederick Ungar, New York, 1954.

21. I. T. Adamson, *Introduction to Field Theory*, University Mathematical Texts, Wiley, New York, 1964.

22. Ian Stewart, *Galois Theory*, Chapman and Hall, London, 1973.

23. Lisl Gaal, *Classical Galois Theory with Examples*, 2nd ed., Chelsea, New York, 1973.

24. Emil Artin, *Geometric Algebra*, Interscience, New York, 1957.

CHAPTER **4**

POLYNOMIALS WITH SYMMETRIC GROUPS

Section 4.1. Background Information

The proof of Abel's Theorem showed that for every $n \geqslant 5$ there exists a polynomial f of degree n over $\mathbf{Q}$ such that *at least one* root of f cannot be expressed in radicals. A natural question is whether a stronger result may actually be true, namely: **For each value of $n \geqslant 5$, does there exist a polynomial f of degree n over Q such that none of the roots of f can be expressed in radicals**? As was pointed out just after the proof of Abel's Theorem, the quintic polynomial used there actually has this stronger property. It follows from Problems 5, 6, and 7 of the same section that for all prime values of $n > 5$, there also exists such a polynomial. In a slightly different vein, if we drop the restriction that the coefficient field be $\mathbf{Q}$, Problem 8 of that section shows that for every $n \geqslant 5$ there always exists some polynomial of degree n, none of whose roots can be obtained by a sequence of radical extensions of the coefficient field.

All of these partial results were obtained by finding polynomials whose Galois groups were S_n, for various values of n. Thus it is natural to attack the previous question by asking: **For what values of n does there exist a**

polynomial over Q whose Galois group is S_n? This question is quite difficult, and our solution will make use of techniques of both an algebraic and an analytic nature.

In this section we shall recall certain facts from calculus concerning infinite series, and we shall use these to develop a classical theorem of algebraic geometry that is basic to our approach. It will also be necessary to develop a rather technical mean value theorem. In the next section, this material will be used to establish a fundamental result, the Irreducibility Theorem of Hilbert. This theorem asserts that if $f(t_1, t_2, \ldots, t_n, x)$ is an irreducible polynomial in $n+1$ variables over $\mathbf{Q}$, then there exist rational values for t_1 through t_n such that the resulting polynomial in x is irreducible over $\mathbf{Q}$. Using this theorem, we shall show in the last section that for every positive integer n, there actually does exist a polynomial over $\mathbf{Q}$ whose group is S_n. If one wishes simply to assume Hilbert's Theorem, it is possible to proceed directly to Section 4.3 without first studying Sections 4.1 and 4.2.

It is supposed that the reader is acquainted with the basic theory of power series in the real domain. Let us recall the relevant facts. For every power series $\sum_{k=0}^{\infty} a_k t^k$ there exists a value R, $0 \leqslant R \leqslant \infty$, called its *radius of convergence*, such that for $|t| < R$ the series converges and for $|t| > R$ it diverges. (There are various possibilities for $|t| = R$, but these do not concern us.) In fact, for $|t| < R$ the original series converges absolutely, meaning that the series $\sum_{k=0}^{\infty} |a_k t^k|$ converges. Absolute convergence at any point always implies convergence. Given two power series with respective radii of convergence R_1 and R_2, we shall refer to those t satisfying both $|t| < R_1$ and $|t| < R_2$ as being "within their common radius of convergence".

Power series in t are simply a generalization of polynomials in t, and they can often be manipulated in an analogous fashion. In particular, within their common radius of convergence, power series can be added, subtracted, and multiplied to give new power series. They can also be divided as long as the denominator can be restricted to a radius within which it is never 0. The rules for such operations are analogous to those for polynomials. A computational example is given in Problem 1. Power series can also be differentiated term by term within their radii of convergence.

An important theorem of which we shall make use is the **Identity Theorem for Power Series**, which says that if two power series have the same values on some interval (or even on any infinite set containing at least one point of accumulation), then their corresponding coefficients must all be identical. In particular, if we have a power series $\sum_{k=0}^{\infty} a_k t^k$ identically equal to 0 on some interval, then each a_k must be 0.

Let $\sum_{k=0}^{\infty} a_k t^k$ have radius of convergence $R > 0$. Since for each individual t, $|t| < R$, we can multiply through by any number and still have a convergent series, it follows that for $0 < |t| < R$ we can also multiply through by monomials of the form At^{-j}. The result is a series of the form

$$b_{-j}t^{-j} + b_{-j+1}t^{-j+1} + \cdots + b_0 + b_1 t + b_2 t^2 + \cdots .$$

It is easy to see that such series can also be added, subtracted, and multiplied in the obvious fashion. (See Problem 2.) They can also be differentiated term by term.

It is necessary to generalize slightly the context within which we shall work. So far we have considered series of the form $\sum_{k=0}^{\infty} a_k t^k$ where the coefficients a_k are real. It is necessary for us to allow complex coefficients. If we have a sequence of complex coefficients a_k and we write them in terms of their real and imaginary parts, $a_k = b_k + ic_k$, then let us make the definition

$$\sum_{k=0}^{\infty} a_k t^k = \sum_{k=0}^{\infty} b_k t^k + i \sum_{k=0}^{\infty} c_k t^k.$$

The minimum of the radii of convergence of the two real power series on the right will be taken as the definition of the radius of convergence of $\sum_{k=0}^{\infty} a_k t^k$. With these conventions, all the previous statements about series carry through immediately to this case. Of course $|a_k|$ must then be interpreted as the modulus of the complex number a_k. (Problem 3 asks for a verification of one such statement for this case. It might be advisable for the reader to think through some of the others.)

A further generalization is possible, namely to the case where t is also allowed to assume complex values. All previous statements still hold in this context, but their proofs do not simply follow from the real case. The theory of such series needs to be constructed from the beginning, although in a manner analogous to the real case. For our work it is completely adequate to consider only real values of t, so we do not depend upon this theory. However, the reader who is comfortable with the more general case may be interested in considering our calculations to take place there, where they are both valid and formally identical.

A complex-valued function f of a real variable t is said to be *analytic at* 0 if it can be represented by a convergent power series,

$$f(t) = \sum_{k=0}^{\infty} a_k t^k,$$

for all t satisfying $|t| < R$, for some $R > 0$. If f is defined on such an interval then we may define a function

$$g(u) = f\left(\frac{1}{u}\right)$$

for all $|u| > 1/R$. We say that g is *analytic at* ∞ if and only if f is analytic at 0. In this case we can express g in the form

$$g(u) = \sum_{k=0}^{\infty} a_k u^{-k}$$

for $|u|$ sufficiently large. Such a series is called a *reciprocal power series*. It is immediate from working in terms of the variable $t = 1/u$ that all our operations on power series with positive radii of convergence are valid for reciprocal power series when $|u|$ is sufficiently large. Similarly, f is said to be *analytic at* t_0, if the function

$$g(u) = f(u + t_0)$$

is analytic at 0. In this case we have an expansion

$$f(t) = \sum_{k=0}^{\infty} a_k u^k = \sum_{k=0}^{\infty} a_k (t - t_0)^k$$

valid in an interval $|t - t_0| < R$ for some $R > 0$.

Leaving power series aside for the moment, there is one last preliminary observation that we shall need. If $f(y)$ is a polynomial in y of degree n and if y_0 is any fixed value of y, then $f(y)$ may be expressed as a polynomial of degree n in the variable $(y - y_0)$. It has the particular form

$$f(y) = f(y_0) + f^{(1)}(y_0)(y - y_0) + \frac{f^{(2)}(y_0)}{2!}(y - y_0)^2 + \cdots + \frac{f^{(n)}(y_0)}{n!}(y - y_0)^n.$$

This is a special case of Taylor's Theorem, but its direct verification in the case of polynomials is very simple (Problem 4).

Now we are ready to begin our development. In what follows it will be completely adequate to think of t as a real variable, although this restriction will not be made explicitly except where it is actually necessary for the validity of the calculations.

Let $f(t, x)$ be a polynomial in two variables over a field **F**. The set C of points (t, x) which are solutions to the equation

$$f(t, x) = 0$$

is called an *affine plane curve*. The nature of such sets is studied extensively in algebraic geometry, and we shall need to derive one of the fundamental classical results (Lemma 36b). By grouping together terms containing the same powers of x, we can write

$$f(t, x) = a_n(t)x^n + a_{n-1}(t)x^{n-1} + \cdots + a_0(t),$$

where each $a_k(t)$ is a polynomial in t over **F** and where $a_n(t)$ is not the zero polynomial. For each fixed value $t = t_0$ such that $a_n(t_0) \neq 0$, there are exactly n roots to the resulting polynomial in x, counted according to multiplicity. If these n roots are distinct, then t_0 is called a *regular value* of t. We shall shortly see that for all t sufficiently close to such a value t_0, the n roots of the equation $f(t, x) = 0$ can be represented by n "root functions" $x_1(t), x_2(t), \ldots, x_n(t)$ which are analytic at t_0. But first it is appropriate to verify that regular values always exist for the particular class of polynomials of interest to us.

Given a polynomial $f(t, x)$ over **F**, as above, we say that it is *irreducible* over **F** if it cannot be written as a product

$f(t, x) = g(t, x)h(t, x)$, where g and h are also polynomials over **F**, each of degree at least one. Most points of affine plane curves corresponding to irreducible polynomials $f(t, x)$ have regular values of t, as we shall presently see. Incidentally, there are actually two senses in which the word "degree" will be used in our discussion. The *degree* of a polynomial in two or more variables, as defined in Section 1.6, is the highest sum of the exponents of the variables among all nonzero terms. In contrast, the *degree in a particular variable* will be taken to mean the highest power of that variable found in the polynomial. Thus, for example, the polynomial $tx^3 + t^4x$ has degree 5, but its degree in x is 3.

LEMMA 36a. *Let $f(t, x)$ be an irreducible polynomial in two variables over the field* **F**. *Then all but a finite number of values of t are regular values.*

Proof. We exclude at the outset those finite number of t values which are roots of $a_n(t)$. Any other value $t = t_0$ for which the g.c.d. of the polynomials in x, $f(t_0, x)$ and $f_x(t_0, x)$, is a nonzero constant must be regular. To see this, recall that the derivative is 0 at a multiple root, in which case the polynomial and its derivative have a common factor and hence a g.c.d. of degree greater than 0.

Given any two polynomials $g(t, x)$ and $h(t, x)$ over **F**, we can write them in the form

$$g(t, x) = b_k x^k + b_{k-1}x^{k-1} + \cdots + b_0,$$

$$h(t, x) = c_m x^m + c_{m-1}x^{m-1} + \cdots + c_0,$$

keeping in mind that the coefficients are polynomial functions of t. From the calculations made in the proof of the division algorithm (Theorem 11), it follows that there are polynomials q and r in x such that

$$g = hq + r$$

and such that r has a lower degree in x than does h. The coefficients of q and r are specified as particular rational combinations of the coefficients of g and h, and so in this case they are quotients of polynomials in t over $\mathbf{F}$, called *rational functions* of t over $\mathbf{F}$. For arbitrary t we now apply the Euclidean algorithm to find the g.c.d. of $f(t, x)$ and $f_x(t, x)$, considered as functions of x. This only involves repeated use of the division algorithm, and so the result has the general form of a polynomial in x with rational functions for its coefficients, say

$$r(t, x) = \frac{p_j(t)}{q_j(t)} x^j + \frac{p_{j-1}(t)}{q_{j-1}(t)} x^{j-1} + \cdots + \frac{p_0(t)}{q_0(t)} .$$

This divides into $f(t, x)$ to give a quotient polynomial of the same type. That is,

$$f(t, x) = r(t, x)s(t, x),$$

where s is a polynomial in x with rational functions of t as its coefficients. By an argument analogous to the proof of Gauss' Lemma (Lemma 12), it is easy to verify that r and s may be modified to new polynomials $\tilde{r}$ and $\tilde{s}$ of the same degree in x but with coefficients that are polynomials in t, so that

$$f(t, x) = \tilde{r}(t, x)\tilde{s}(t, x).$$

(See Problem 5.) By the irreducibility of f, $\tilde{r}(t, x)$ must be a constant. Since r has the same degree in x, we have

$$r(t, x) = \frac{p_0(t)}{q_0(t)} ,$$

where p_0 is not identically 0, by the stopping criterion of the Euclidean algorithm. As in the proof of the latter, we can write

$$S(t,x)f(t,x)+T(t,x)f_x(t,x)=\frac{p_0(t)}{q_0(t)},$$

where S and T are polynomials in x with rational functions of t for their coefficients. Excluding the finite set of values of t which are roots of any of the denominators or of $p_0(t)$, we conclude that f and f_x cannot have any other common roots. For at such points the left side would be zero and the right side would not. Thus for all but a finite set of values, each value $t = t_0$ causes the g.c.d. of $f(t_0, x)$ and $f_x(t_0, x)$ to be a nonzero constant. As noted at the outset, this completes the proof. ∎

Careful scrutiny of the proof of the previous lemma shows that it still applies if the hypothesis on $f(t, x)$ is weakened somewhat. In particular, it is sufficient to assume that $f(t, x)$ is *irreducible in* x, meaning that we cannot write $f(t, x) = g(t, x)h(t, x)$, with g and h each polynomials of degree at least one in x. For example, the polynomial over **Q** given by $f(t, x) = tx^2 + t$ is irreducible in x even though it is not irreducible in the general sense. Several of our subsequent results may also be given with similar weakened hypotheses, but this would add unnecessary complexity both to their statements and their proofs. Since this generalization is not relevant to the application we have in mind, it will not be pursued further.

It is also appropriate to remark at this point that the similarity in the arithmetic of polynomials in x with coefficients in a field **F** and polynomials in x with coefficients that are rational functions in t suggests the possibility of extending the definition of field so as to include both cases. For example, the set of rational functions in t over $\mathbf{F}$, say, is closed under addition, subtraction, multiplication, and division (except by 0), and thus it has the same basic arithmetic structure as the sets of numbers we have called fields. This generalization of the notion of a field is very powerful and valuable, and it represents the point of view one would encounter in the subject of abstract algebra. For our purposes it is more convenient to deal with the concrete notion of a field of complex numbers, and so the abstract point of view will not be developed here.

Next we see how the points of an affine plane curve C can be represented in the neighborhood of a regular value of t.

LEMMA 36b. *Let t_0 be a regular value of the affine plane curve given by $f(t, x) = 0$. Then there exist n functions $x_1(t), x_2(t), \ldots, x_n(t)$ which are analytic at t_0 and such that for every t in some neighborhood of t_0, they represent the complete set of solutions to $f(t, x) = 0$.*

Proof. Without loss of generality we may take $t_0 = 0$, for the transformation $g(t, x) = f(t + t_0, x)$ may be used to put the problem in this form. Since 0 is a regular point, there are n distinct roots, which we may denote by $x_1(0), x_2(0), \ldots, x_n(0)$. We shall show that each of these gives rise to an analytic function $x_i(t)$ as described in the theorem.

Let $x(t)$ denote any one of the sought root functions $x_i(t)$. Without loss of generality we may also assume that $x(0) = 0$, for the transformation $g(t, x) = f(t, x + x(0))$ may be used to obtain this form. Consequently, we now seek a power series

$$x(t) = \sum_{k=1}^{\infty} b_k t^k$$

which converges on some set $|t| < R$, $R > 0$, within which

$$f(t, x(t)) = 0.$$

Since by our assumptions $f(0, 0) = 0$, we may write f in the form

$$f(t, x) = a_{1,0}t + a_{0,1}x + \sum_{i+j \geqslant 2} a_{ij}t^i x^j.$$

The sum on the right is a finite sum because f is a polynomial. Since $t = 0$ is a regular value, $f_x(0, 0) = a_{0,1} \neq 0$, for otherwise $x = 0$ would be a multiple root. Consequently we may divide f by $-a_{0,1}$ without changing the solutions. Equivalently, without loss of generality we may take $a_{0,1} = -1$.

Our proof has two main parts. First we determine a unique set of coefficients b_k, $k \geqslant 1$, in order for $x(t)$ to be a candidate for a solution to $f(t, x(t)) = 0$. Then we show that for this choice of the b_k's, the resulting series actually has a positive radius of convergence and does satisfy the equation.

From the equation

$$f(t, x) = a_{1,0}t - x + \sum_{i+j \geqslant 2} a_{ij}t^i x^j$$

it follows that

$$f_x(t, x) = -1 + g(t, x),$$

where each term of g has degree at least one. By

substituting $x = \sum_{k=1}^{\infty} b_k t^k$ into the equation $f(t, x) = 0$, it follows that we must have $b_1 = a_{1,0}$. For the only terms involving t raised to the first power are $a_{1,0}t$ and $-b_1 t$; by the Identity Theorem for Power Series they must add to 0. Now we obtain a recursive definition for the b_k's, $k \geqslant 2$, by taking a Taylor expansion of the polynomial in $y, f(t, y)$, around the point $y_0 = \sum_{i=1}^{k-1} b_i t^i$. For the value $y = \sum_{i=1}^{\infty} b_i t^i$ we obtain

$$0 = f\left(t, \sum_{i=1}^{\infty} b_i t^i\right)$$

$$= f\left(t, \sum_{i=1}^{k-1} b_i t^i\right) + \left[-1 + g\left(t, \sum_{i=1}^{k-1} b_i t^i\right)\right] \sum_{i=k}^{\infty} b_i t^i$$

$$+ \text{ terms of degree at least } 2k.$$

By the Identity Theorem for Power Series, the coefficient of each power of t must be 0. In particular, the coefficient of t^k is $-b_k$ plus the coefficient of t^k in the expansion of the polynomial $f(t, \sum_{i=1}^{k-1} b_i t^i)$. Therefore, b_k must equal this coefficient. With this rule and the initial value $b_1 = a_{1,0}$, the entire sequence is determined.

Let us make two further observations from this calculation. First, if the resulting series $\sum_{k=1}^{\infty} b_k t^k$ does have a positive radius of convergence, then $0 = f(t, \sum_{k=1}^{\infty} b_k t^k)$, for this latter is a power series every one of whose coefficients is 0. Second, it is easy to see inductively that each b_k can be written in the abbreviated form $p_k(a_{ij})$, where p_k is a polynomial in several variables, with positive integral coefficients, evaluated at the original coefficients a_{ij}.

Now we proceed to show that $\sum_{k=1}^{\infty} b_k t^k$ has a positive radius of convergence. We shall do this by constructing a power series $\sum_{k=1}^{\infty} A_k t^k$ that has a positive radius of convergence R and such that for each k, $A_k \geqslant |b_k|$. Thus for $|t| < R$, we will have

$$\sum_{k=1}^{\infty} |b_k t^k| \leqslant \sum_{k=1}^{\infty} |A_k t^k| < \infty,$$

from which follows the convergence of $\sum_{k=1}^{\infty} b_k t^k$.

Let A be any number greater than or equal to every single $|a_{ij}|$. Following our previous calculation, but with each a_{ij} replaced by A in the original equation, we see that if this new equation has an analytic solution at 0, the solution must have the form $\sum_{k=1}^{\infty} A_k t^k$ with $A_k = p_k(A)$ for each k. Thus if we can demonstrate independently that this equation does have an analytic solution at 0, the argument will be complete. The equation is

$$0 = At - x + A \sum_{i+j \geqslant 2} t^i x^j$$

and it is actually quite easy to solve. Using the formula for the sum of a geometric progression, we obtain

$$\begin{aligned} 0 &= At - x + At^0 \sum_{j=2}^{\infty} x^j + At^1 \sum_{j=1}^{\infty} x^j + A \sum_{i=2}^{\infty} t^i \left(\sum_{j=0}^{\infty} x^j \right) \\ &= At - x + A \frac{x^2}{1-x} + At \frac{x}{1-x} \\ &\quad + A \left(\frac{t^2}{1-t} \right) \left(\frac{1}{1-x} \right). \end{aligned}$$

Manipulating algebraically, we have

$$0 = At - Atx - x + x^2 + Ax^2 + Atx + A\left(\frac{t^2}{1-t}\right)$$

$$= [A+1]x^2 + [-1]x + \left[A\frac{t}{1-t}\right],$$

which is a quadratic equation in x. The solution satisfying $x(0) = 0$ is given by

$$x = \frac{1 - \sqrt{1 - 4(A+1)At/(1-t)}}{2(A+1)}.$$

The radical may be expressed as a quotient of functions of the form $\sqrt{1-Kt}$. From calculus (see also Problem 6) we know that $\sqrt{1-u}$ may be expressed in a power series in u for $|u| < 1$, from which it is clear that $\sqrt{1-Kt}$ is also analytic at 0. By the validity of the arithmetic operations on power series, we conclude that x is analytic at 0.

We have thus shown that to each $x_i(0)$ there corresponds an analytic root function $x_i(t) = x_i(0) + \sum_{k=1}^{\infty} b_k t^k$. By their continuity at 0 these functions are all distinct in some neighborhood of $t = 0$ and thus they represent the complete set of solutions in such a neighborhood. ∎

The last preliminary result that we shall need is a fairly simple observation from interpolation theory, although its analytic statement appears somewhat complicated. Given $m+1$ increasing values of the real variable t, $t_0 < t_1 < t_2 < \cdots < t_m$, let us denote by V_m the Vandermonde determinant

$$V_m = \det \begin{bmatrix} 1 & t_0 & t_0^2 & \cdots & t_0^{m-1} & t_0^m \\ 1 & t_1 & t_1^2 & \cdots & t_1^{m-1} & t_1^m \\ \vdots & & & & & \\ 1 & t_m & t_m^2 & \cdots & t_m^{m-1} & t_m^m \end{bmatrix}.$$

For a given function $z(t)$, we denote by W_m the determinant obtained by changing the last column of V_m as follows:

$$W_m = \det \begin{bmatrix} 1 & t_0 & t_0^2 & \cdots & t_0^{m-1} & z(t_0) \\ 1 & t_1 & t_1^2 & \cdots & t_1^{m-1} & z(t_1) \\ \vdots & & & & & \\ 1 & t_m & t_m^2 & \cdots & t_m^{m-1} & z(t_m) \end{bmatrix}.$$

With this notation, we may state the lemma:

LEMMA 36c. *If $z(t)$ is m times differentiable on the interval $t_0 \leqslant t \leqslant t_m$, then there exists some value $\bar{t}$, $t_0 < \bar{t} < t_m$, such that*

$$\frac{z^{(m)}(\bar{t})}{m!} = \frac{W_m}{V_m}.$$

Proof. First we prove the more general result that if $y(t)$ is any m times differentiable function on the same interval, such that $y(t_i) = z(t_i)$ for every i, $0 \leqslant i \leqslant m$, then $y^{(m)}(\bar{t}) = z^{(m)}(\bar{t})$ for some $\bar{t}$, $t_0 < \bar{t} < t_m$. For the Mean Value Theorem, applied to the function $y - z$ on each interval $[t_i, t_{i+1}]$, implies that $y^{(1)}(t)$ and $z^{(1)}(t)$ agree on m distinct points. By the same argument, $y^{(2)}(t)$ and $z^{(2)}(t)$ agree on $m-1$ points. Continuing in this fashion we obtain the desired conclusion.

The lemma follows upon taking for $y(t)$ the polynomial of degree m which interpolates to the values of $z(t)$ at each t_i. In particular, to construct such a polynomial $y(t) = a_0 + a_1 t + \cdots + a_m t^m$ we simply need to solve the following linear system for the a_j's:

$$a_0 + a_1 t_0 + \cdots + a_{m-1} t_0^{m-1} + a_m t_0^m = z(t_0)$$

$$a_0 + a_1 t_1 + \cdots + a_{m-1} t_1^{m-1} + a_m t_1^m = z(t_1)$$

$$\vdots$$

$$a_0 + a_1 t_m + \cdots + a_{m-1} t_m^{m-1} + a_m t_m^m = z(t_m).$$

By Cramer's rule,

$$a_m = \frac{W_m}{V_m}.$$

(V_m is not 0 since $V_m = \prod_{i>j}(t_i - t_j)$ and the t_i's are distinct.) Since $y^{(m)}(t)$ has the constant value $m! a_m$, the equation $z^{(m)}(\bar{t}) = y^{(m)}(\bar{t})$ yields the statement of the lemma. ∎

PROBLEMS

1. The function e^t is sometimes defined by the power series $\sum_{n=0}^{\infty} t^n / n!$. Show that this series converges for all t. Then use multiplication of series to prove the law of exponents $e^a e^b = e^{a+b}$.

2. On the basis of the fact that power series can be

added term by term within their common radius of convergence, show that two series of the form $f(t) = \sum_{k=-j}^{\infty} a_k t^k$ and $g(t) = \sum_{k=-j}^{\infty} b_k t^k$ can also be added term by term. Here $j > 0$ and f and g both converge when $0 < |t| < R$.

3. On the basis of the definition of series of the form $\sum_{k=0}^{\infty} a_k t^k$ in the case when the a_k's may not be real, justify the fact that two such series may be multiplied in the usual way within their common radius of convergence.

4. Prove Taylor's Theorem for the case of a polynomial: If f is a polynomial of degree n in y, then

$$f(y) = f(y_0) + f^{(1)}(y_0)(y - y_0) + \frac{f^{(2)}(y_0)}{2!}(y - y_0)^2$$

$$+ \cdots + \frac{f^{(n)}(y_0)}{n!}(y - y_0)^n.$$

Here y_0 is an arbitrary fixed value of y.

5. Prove the following version of Gauss' Lemma: If $f(t, x)$ is a polynomial in two variables over $\mathbf{F}$ and if f can be written as a product $f(t, x) = r(t, x)s(t, x)$ where r and s are polynomials in x whose coefficients are rational functions of t over $\mathbf{F}$, then in fact f may be written $f(t, x) = \tilde{r}(t, x)\tilde{s}(t, x)$ where $\tilde{r}$ and $\tilde{s}$ are polynomials over $\mathbf{F}$ in two variables. Furthermore, $\tilde{r}$ and r have the same degree in x, as do $\tilde{s}$ and s.

6. Prove that the function $\sqrt{1 - t}$ is analytic at 0.

*7. Let $f(t)$ be analytic at 0 and suppose $f(0) > 0$. Show that f has an analytic square root at 0; that is, show that there is a function $g(t)$ defined in some neighborhood of 0 such that $g(t)$ is analytic at 0 and $[g(t)]^2 = f(t)$.

8. Consider the algebraic curve C defined by $f(t, x) = 0$. A *real point* of C is a point where both t and x

are real. For example, the real points of $t^2 - x^2 = 0$ form two lines in the real t, x-plane. The curve $t^2 + x^2 = 0$ has only (0, 0) for a real point. A real point is *isolated* if it has some neighborhood containing no other real points, as in the second example. Show that if $f(t, x)$ has real coefficients, then for every regular value of t, any corresponding real point cannot be isolated.

9. Suppose that $y = g(t)$ is a polynomial of degree m over **C**. If there exist $m + 1$ values $t_i \in \mathbf{Q}$ at which $g(t_i) \in \mathbf{Q}$, show that the coefficients of $g(t)$ must actually be in **Q**.

Section 4.2. Hilbert's Irreducibility Theorem

The theorem to be proved in this section may be regarded as the fundamental tool in our treatment of polynomials over **Q** with symmetric groups. We shall first prove a special case of the theorem as Lemma 36d, and by using this it will not be difficult to obtain the general result.

Suppose that $f(t, x)$ is an irreducible polynomial over **Q** in two variables. It is useful to observe that f may be transformed into another polynomial over **Q** that has certain convenient properties. Let $f(t, x)$ be written in the form

$$f(t, x) = a_n(t)x^n + a_{n-1}(t)x^{n-1} + \cdots + a_0(t),$$

where each $a_i(t)$ is a polynomial over **Q**, and suppose d is the highest degree of any of these polynomials $a_i(t)$. Since f is irreducible, all but a finite number of t values are regular, and so we may pick a regular value t_0 which is rational. The function $g(t, x)$ defined by

$$g(t, x) = t^d f\left(t_0 + \frac{1}{t}, x\right)$$

is readily seen to be a polynomial over $\mathbf{Q}$. It has three important properties. First, it has n distinct root functions that are analytic at infinity; we thus say that g has a *regular value at infinity*. To see that this is the case, note that for values of the variable $t_0 + (1/t)$ sufficiently close to t_0, the solutions to the equation $f(t_0 + (1/t), x) = 0$ may be expressed as power series in the variable $t_0 + (1/t) - t_0$, which is simply $1/t$. Second, $g(t, x)$ is also irreducible. For if we could write

$$g(t, x) = r(t, x)s(t, x)$$

with r and s each of degree at least one, we could also write

$$\begin{aligned} f(t, x) &= (t - t_0)^d g\left(\frac{1}{t - t_0}, x\right) \\ &= (t - t_0)^d r\left(\frac{1}{t - t_0}, x\right) s\left(\frac{1}{t - t_0}, x\right). \end{aligned}$$

Since the sum of the highest degrees of the coefficient polynomials in r and s must equal d, it follows that the above expression would imply a nontrivial factorization of $f(t, x)$, contrary to our hypothesis. Third, if $t_1 \neq t_0$ is a rational value at which the polynomial $g(t_1, x)$ is irreducible in $\mathbf{Q}[x]$, then the corresponding rational value $\tilde{t}_1 = t_0 + (1/t_1)$ is such that $f(\tilde{t}_1, x)$ is also irreducible in $\mathbf{Q}[x]$. This follows from the above equations upon setting $t = t_1$.

We are now ready to prove Hilbert's theorem for the case of two variables:

LEMMA 36d. *Let $f(t, x)$ be an irreducible polynomial in two variables over $\mathbf{Q}$. Then there exist an infinite number of rational values t_0 such that $f(t_0, x)$ is irreducible in $\mathbf{Q}[x]$.*

Furthermore, if $f_1(t, x), \ldots, f_M(t, x)$ are M such polynomials, then there are an infinite number of rational values t_0 such that $f_1(t_0, x), \ldots, f_M(t_0, x)$ are all irreducible in $\mathbf{Q}[x]$.

Proof. As usual, we write $f(t, x)$ in the form

$$f(t, x) = a_n(t)x^n + a_{n-1}(t)x^{n-1} + \cdots + a_0(t).$$

On the basis of the comments preceding the lemma, we can assume without loss of generality that f has a regular value at infinity. Thus for all t greater than or equal to some T_0, there are n distinct root functions $x_1(t)$, $x_2(t)$, $\ldots, x_n(t)$ and each may be expressed as a reciprocal power series. For $t \geqslant T_0$, we also know $a_n(t) \neq 0$. For each fixed t, $f(t, x)$ can certainly be factored in $\mathbf{C}[x]$ as a product of linear factors; in particular we can write

$$f(t, x) = a_n(t) \prod_{i=1}^{n} [x - x_i(t)]$$

for all $t \geqslant T_0$ and all x. Because $f(t, x)$ is irreducible over $\mathbf{Q}$, the product $\prod_{i=1}^{n}[x - x_i(t)]$ cannot be written as a product of nontrivial polynomials over $\mathbf{Q}$. Consequently, if S is any nonempty proper subset of the set $\{1, 2, \ldots, n\}$, then the factorization

$$\prod_{i=1}^{n} [x - x_i(t)] = \prod_{i \in S} [x - x_i(t)] \cdot \prod_{i \notin S} [x - x_i(t)]$$

does not represent a factorization into polynomials in two variables over $\mathbf{Q}$. Each of the factors is a polynomial in x, of course, and so at least one of the coefficients is not a polynomial in t over $\mathbf{Q}$. But even further, one of the coefficients is not even a rational function in t over $\mathbf{Q}$, because if they all were rational functions, our recent version of Gauss' Lemma (Problem 5, Section 4.1) would yield a nontrivial factorization of $f(t, x)$. There are $2^n - 2$ such sets S and thus $N = 2^{n-1} - 1$ distinct factorizations of the above type. For each such factorization, we choose one coefficient function that is not a rational function of t over $\mathbf{Q}$. In this way we obtain a list of functions $y_1(t)$, $y_2(t), \ldots, y_N(t)$. By converse reasoning, if $t_0 \geqslant T_0$ is a rational value of t such that each of the numbers $y_1(t_0)$, $y_2(t_0), \ldots, y_N(t_0)$ is irrational, then $f(t_0, x)$ is irreducible in $\mathbf{Q}[x]$. To see this note that $a_n(t_0)$ is a nonzero rational, and thus a factorization of $f(t_0, x)$ in $\mathbf{Q}[x]$ would imply a factorization of $\prod_{i=1}^n [x - x_i(t_0)]$ in $\mathbf{Q}[x]$. But without loss of generality we need only consider factors of the form $\prod_{i \in S}[x - x_i(t)]$ and $\prod_{i \notin S}[x - x_i(t)]$, and these possibilities are all excluded by the choice of t_0.

Thus the first part of the lemma will be proved if we can find an infinite number of rational values $t_0 \geqslant T_0$ such that all the numbers $y_i(t_0)$ are simultaneously irrational. We shall actually manage to do this with integers t_0. Let us begin by taking an arbitrary $y_i(t)$, henceforth denoted $y(t)$, and studying how many integers $t_0 \geqslant T$ have the property that $y(t_0)$ is rational. In a certain sense this set is small.

The function $y(t)$ is obtained from some subset of the root functions $x_i(t)$ by no more than addition, subtraction, and multiplication. From this we deduce two things: $y(t)$ may be represented by a reciprocal power series for $t \geqslant T_0$, and for each such t that is rational, $y(t)$ is algebraic over $\mathbf{Q}$. This latter fact follows from Theorem 8. By reviewing the proof of this same theorem, we further see that we can construct a polynomial in y with coefficients which are rational functions of t over $\mathbf{Q}$, such that for each t, $y(t)$ is a root. Thus we have that $y(t)$ is a root of an equation of the form

$$d_j(t)y^j + d_{j-1}(t)y^{j-1} + \cdots + d_0(t) = 0.$$

Without loss of generality we may take the coefficient functions to be polynomials, and then with integral coefficients, for multiplication through by a suitable polynomial over $\mathbf{Q}$ would convert rational d_i's to this form. Now if we multiply this equation through by $[d_j(t)]^{j-1}$ and make a new function $z(t) = d_j(t)y(t)$, we have $z(t)$ as a solution to a polynomial equation

$$z^j + b_{j-1}(t)z^{j-1} + \cdots + b_0(t) = 0$$

where the b_i's are polynomials in t with integral coefficients. There is a real advantage to working with the function $z(t)$ rather than $y(t)$. If t_0 is an integer for which $y(t_0)$ is rational, then $z(t_0)$ has to be an integer. To see this, first note that $z(t_0)$ is rational and each $b_i(t_0)$ is integral. By the Rational Roots Theorem (Problem 2, Section 1.5), the only possible rational values of $z(t_0)$ are integers.

Thus we now want to show that the set of integers $t_0 \geqslant T_0$ for which $z(t_0)$ is an integer is 'small'. From the definition of $z(t)$ as the product of a polynomial and a reciprocal power series, we can write it in the form

$$z(t) = c_k t^k + \cdots + c_1 t + c_0 + c_{-1}t^{-1} + \cdots .$$

If z is actually a polynomial, then one of its coefficients must be irrational, for otherwise $y(t) = z(t)/d(t)$ would be a rational function over $\mathbf{Q}$, contrary to our hypothesis. But if z is a polynomial with an irrational coefficient, then by Problem 9 of Section 4.1 there can only be a finite number of integers t_0 for which $z(t_0)$ is an integer. In this case we can pick a $T_1 \geqslant T_0$ such that for all integers $t_0 \geqslant T_1$, $z(t_0)$ is irrational.

If any coefficient c_i is not real, then it is also the case that there can be only a finite number of integers $t_0 \geqslant T_0$ for which $z(t_0)$ is an integer. To see this, let i be the largest subscript of such a c_i. Then $\lim_{t\to\infty} \operatorname{Im}\{z(t)/t^i\} = \operatorname{Im} c_i$, and so for all t greater than or equal to some $T_1 \geqslant T_0$, $z(t)$ is not even real.

The final case then is where all the c_i's are real and where at least one c_i with a negative subscript is not 0. In this case, by differentiating $z(t)$ a sufficient number of times we can eliminate all nonnegative powers of t. The result can be written

$$z^{(m)}(t) = \frac{p}{t^q} + \cdots$$

where p is a nonzero constant, q is a positive integer (actually $\geqslant m+1$), and the dots represent terms with higher powers of $1/t$. Since $\lim_{t\to\infty} t^q z^{(m)}(t) = p$, there exists a $T_1 \geqslant T_0$ such that for all $t \geqslant T_1$, $0 < |z^{(m)}(t)| \leqslant 2|p|/t^q$.

It is now time to make use of Lemma 36c. If there are an infinite number of integers $t \geqslant T_1$ such that $z(t)$ is also an integer, let us determine how much they must be spread out. Let $t_0 < t_1 < \cdots < t_m$ be $m+1$ such integers. Using the notation of Lemma 36c, we see that W_m is a nonzero integer and so $|W_m| \geqslant 1$. Thus we have

$$\frac{2|p|}{m!\, t_0^q} \geqslant \frac{2|p|}{m!\, \bar{t}^q} \geqslant \frac{|z^{(m)}(\bar{t})|}{m!} \geqslant \frac{1}{|V_m|}.$$

Therefore we also have

$$\frac{m!}{2|p|} t_0^q \leqslant |V_m| = \prod_{i>j} (t_i - t_j) < (t_m - t_0)^{m(m+1)/2}.$$

This implies that there are positive constants α and β such that $t_m - t_0 > \alpha t_0^\beta$. Since the right side increases with t_0, we see that for larger values of t, these t_i's become more spread out. In particular, choose a number $T_2 \geqslant T_1$ such that $\alpha T_2^\beta \geqslant Nm$. (Recall that N is the number of functions $y_j(t)$.) Then any string of $Nm + 1$ consecutive integers greater than or equal to T_2 contains at most m integers t_i for which $y(t_i)$ is rational.

Now we shall demonstrate the existence of integers t such that for all j, $y_j(t)$ is irrational. For those y_j's which only have a finite number of integral values t such that $y_j(t)$ is rational, choose a number T_2 larger than all of these values. For the other y_j's, pick a value of m that will simultaneously serve in all cases. The result of the above computation yields a value T_2. Let T^* be the maximum of all these T_2's. Then any string of $Nm + 1$ consecutive integers greater than T^* contains at most m integers t_i yielding rational values for any one function y_j. Since there are N functions, there is at least one point in each such set at which the functions are simultaneously irrational. This then completes the proof for a single function f. In the case of M functions f_i, we simply work with the union of the y_i's corresponding to each one. Letting the total number now be represented by N, the above argument applies and gives the desired conclusion. ∎

We have proved Hilbert's Irreducibility Theorem for the case of a polynomial in two variables, and we want to develop the analog for polynomials in $n + 1$ variables. To this end we shall use a transformation, called **Kronecker's specialization**, for converting a polynomial in $n + 1$ variables to a polynomial in two variables such that the relevant irreducibility properties are essentially preserved.

Let $f(u_0, u_1, \ldots, u_n)$ be a polynomial over $\mathbf{Q}$ in $n+1$ variables. In what follows, let us think of it as a polynomial in the n variables $u_1, u_2, \ldots, u_n$ which has coefficients that are elements of $\mathbf{Q}[u_0]$. Each term consists of some polynomial in u_0 over $\mathbf{Q}$ multiplied by an expression of the form $u_1^{i_1}u_2^{i_2} \cdots u_n^{i_n}$. Let d be a number such that every one of the exponents $i_1, \ldots, i_n$ for every one of the terms is less than d. Fixing this number d, let P_d be the set of all polynomials over $\mathbf{Q}$ in $u_0, u_1, \ldots, u_n$ having this same property. Then for any polynomial $g(u_0, u_1, \ldots, u_n)$ in P_d we define a new polynomial $\hat{g}(u_0, y)$ by the equation

$$\hat{g}(u_0, y) = g\left(u_0, y, y^d, y^{d^2}, \ldots, y^{d^{n-1}}\right).$$

Thus $\hat{g}$ is a polynomial in two variables u_0 and y. Considered as a polynomial in y with coefficients which are elements of $\mathbf{Q}[u_0]$, the degree of $\hat{g}$ in y is less than or equal to $d^n - 1$. To see this note that under our correspondence a monomial $u_1^{i_1}u_2^{i_2} \cdots u_n^{i_n}$ is associated with a term $y^{i_1 + di_2 + d^2 i_3 + \cdots d^{n-1} i_n}$. The exponent of y has its maximum value $d^n - 1$ when each i_j equals $d-1$. The set of all polynomials over $\mathbf{Q}$ in u_0 and y with degree at most $d^n - 1$ in y will be denoted K_d.

If we replace u_0 by a rational constant α in the original polynomial g, and if we define the transform of the resulting polynomial in the obvious way to be $g(\alpha, y, y^d, y^{d^2}, \ldots, y^{d^{n-1}})$, it is clear that this is the same polynomial in y that would be obtained by first transforming g and then substituting $u_0 = \alpha$. This polynomial is of course $\hat{g}(\alpha, y)$. The properties to be explained below hold both for the case when u_0 is a variable and when u_0 has been given a specific rational value α, and even the notation P_d and K_d will be understood to refer to both cases.

The mapping from P_d to K_d given by $g \to \hat{g}$ is easily seen to be a one-to-one correspondence; that is, it is both one-to-one and onto (Problem 1). Furthermore, if a polynomial $f \in P_d$ can be factored over **Q**, so that we can write $f = gh$ for $g, h \in P_d$, then $\hat{f}$ can also be factored:

$$\hat{f} = \widehat{(gh)} = \hat{g}\hat{h}.$$

(Problem 2.) Each of g and h has degree at least one in some one of the variables $u_1, u_2, \ldots, u_n$ if and only if $\hat{g}$ and $\hat{h}$ each have degree at least one in y.

The converse statements do not hold: $\hat{f}$ may be reducible even if f is not. For example, suppose $\hat{f}$ may be written

$$\hat{f} = GH$$

for $G, H \in K_d$. It is certainly true that there exist $g, h \in P_d$ such that $\hat{g} = G$ and $\hat{h} = H$. However, it is possible that the product gh may not be in P_d, in which case we are not led to a factorization of f. For example, consider the polynomial $f(u_1, u_2) = u_1^2 + u_2^2$, which for simplicity does not even contain the variable u_0. With the value $d = 3$, we obtain the transform $\hat{f}(y) = y^2 + (y^3)^2 = y^2 + y^6$. This is obviously reducible over **Q**, whereas the original polynomial $f(u_1, u_2)$ is not. To see what happens when we factor $\hat{f}$ into a product GH and then construct the corresponding g and h in P_d, suppose we look at the factorization $\hat{f} = GH$ with $G(y) = y$ and $H(y) = y + y^5$. By the definition of our transformation, it is easy to see that $\hat{g} = G$ for $g(u_1, u_2) = u_1$, and $\hat{h} = H$ for $h(u_1, u_2) = u_1 + u_1^2 u_2$. Now g and h are both in P_d, but their product

$$(gh)(u_1, u_2) = u_1^2 + u_1^3 u_2$$

is not, since the exponent of u_1 in the second term is not less than d. Thus we cannot have the factorization $f = gh$. This same phenomenon occurs with the other factorization of $\hat{f}$, namely, $\hat{f}(y) = y^2(1 + y^4)$. (See Problem 3.) These observations lead us to formulate the following precise relationship between the reducibility of $\hat{f}$ and f, called **Kronecker's criterion:**

LEMMA 36e. *In the terminology of the preceding discussion, f is irreducible as a polynomial in $u_1, u_2, \ldots, u_n$ if and only if every nontrivial factorization of its transform, $\hat{f} = GH$, leads back to a product gh which is not in P_d, that is, which contains some one of the variables $u_1, u_2, \ldots, u_n$ raised to a power greater than or equal to d.*

Proof. If f is reducible, say $f = gh$, then as noted above $\hat{f} = \hat{g}\hat{h}$, which implies that $\hat{f}$ is reducible. On the other hand, if f is irreducible, then any factorization $\hat{f} = GH = \hat{g}\hat{h}$ must necessitate that $gh \notin P_d$, for if $gh \in P_d$, the fact that the transform is one-to-one would imply that $f = gh$, contradicting the irreducibility of f. ∎

Let us now use these ideas to complete the proof of the **Hilbert Irreducibility Theorem:**

THEOREM 36. *Let $f(t_1, t_2, \ldots, t_n, x)$ be an irreducible polynomial over $\mathbf{Q}$ in the $n + 1$ variables $t_1, t_2, \ldots, t_n, x$. Then there exist an infinite number of sets of rational values $\alpha_1, \alpha_2, \ldots, \alpha_n$ such that $f(\alpha_1, \alpha_2, \ldots, \alpha_n, x)$ is irreducible in $\mathbf{Q}[x]$.*

Proof. The proof will be by induction on n. When $n = 1$, the theorem reduces to Lemma 36d, the case of Hilbert's theorem for two variables. The induction step consists of reducing the case of n to the case of $n-1$ by showing that there are an infinite number of rational values α such that $f(\alpha, t_2, \ldots, t_n, x)$ is irreducible over $\mathbf{Q}$. For this step it is convenient to relabel the variables as $u_0, u_1, \ldots, u_n$ and to adopt the notation of the previous discussion.

Thus we have that $f(u_0, u_1, \ldots, u_n)$ is irreducible. If it happens that $\hat{f}(u_0, y)$ is irreducible over $\mathbf{Q}$, then Lemma 36d implies the existence of an infinite number of rational values α such that $\hat{f}(\alpha, y)$ is irreducible in $\mathbf{Q}[y]$. By Lemma 36e, for each such value α the polynomial $f(\alpha, u_1, \ldots, u_n)$ is irreducible, which is what we want.

If $\hat{f}(u_0, y)$ is reducible, then a more complicated analysis is necessary. In particular, by repeatedly factoring over $\mathbf{Q}$ until this is no longer possible, we can write

$$\hat{f}(u_0, y) = \prod_i G_i(u_0, y)$$

where each G_i is irreducible over $\mathbf{Q}$. By Lemma 36d, there is an infinite set $\mathcal{Q}$ of rational values α for which all the polynomials $G_i(\alpha, y)$ are simultaneously irreducible in $\mathbf{Q}[y]$. We claim that for all but a finite number of these values α, the polynomial $f(\alpha, u_1, \ldots, u_n)$ is irreducible as a polynomial in n variables. The crucial tool will be Lemma 36e.

There are only a finite number of essentially different ways in which $\hat{f}(\alpha, y)$ might be factorable over $\mathbf{Q}$ into two nontrivial factors. To see this, suppose we have a factorization

$$\hat{f}(\alpha, y) = A(y)B(y),$$

where A and B are polynomials in $\mathbf{Q}[y]$. Since we also have

$$\hat{f}(\alpha, y) = \prod_i G_i(\alpha, y),$$

and since each $G_i(\alpha, y)$ is irreducible, some of the G_i's are factors of $A(y)$ and the others are factors of $B(y)$ (Theorem 14). That is, we can write

$$A(y) = a \prod_{i \in S} G_i(\alpha, y),$$

$$B(y) = a^{-1} \prod_{i \in T} G_i(\alpha, y),$$

for some rational number a and some disjoint sets S and T whose union contains all the subscripts i. Without loss of generality we may take $a = 1$ by dividing the original $A(y)$ by a and multiplying the original $B(y)$ by a. Consider the corresponding factorization of $\hat{f}(u_0, y)$:

$$\hat{f}(u_0, y) = \left(\prod_{i \in S} G_i(u_0, y)\right)\left(\prod_{i \in T} G_i(u_0, y)\right)$$

and call the two factors in parentheses on the right $Q(u_0, y)$ and $R(u_0, y)$ respectively. Since both Q and R are in K_d, there must exist polynomials $q(u_0, u_1, \ldots, u_n)$ and $r(u_0, u_1, \ldots, u_n)$ in P_d such that $\hat{q} = Q$ and $\hat{r} = R$. Note that $Q(\alpha, y) = A(y)$ and $R(\alpha, y) = B(y)$.

Since $f(u_0, u_1, \ldots, u_n)$ is irreducible, the product $q(u_0, u_1, \ldots, u_n)r(u_0, u_1, \ldots, u_n)$ cannot be in P_d, by Lemma 36e, and so some term has too high a power of at least one of the variables u_1 through u_n. The coefficient of this term is a polynomial in u_0, and so long as α is not one of the roots of this polynomial, the product $q(\alpha, u_1, \ldots, u_n)$ $r(\alpha, u_1, \ldots, u_n)$ also will not be in P_d. If now for each of the finite number of groupings of the subscripts into sets S and T we remove from the set $\mathcal{Q}$ that finite set of α values just described, then all the remaining values of α will guarantee, by Lemma 36e, that $f(\alpha, u_1, \ldots, u_n)$ is irreducible. As noted earlier, this then completes the proof. ∎

PROBLEMS

1. Verify that Kronecker's specialization gives a one-to-one correspondence between the sets P_d and K_d.

2. If g, h, and gh are all in P_d, show that $\widehat{(gh)} = \hat{g}\hat{h}$.

3. In the example immediately preceding Lemma 36e, verify that the factorization $\hat{f}(y) = y^2(1 + y^4)$ leads back to a product gh which is not in P_d.

4. This problem refers to the proof of Theorem 36. Define $g_i(u_0, u_1, \ldots, u_n)$ by the requirement that $\hat{g}_i(u_0, y) = G_i(u_0, y)$. Does it follow that $q(u_0, u_1, \ldots, u_n) = \prod_{i \in S} g_i(u_0, u_1, \ldots, u_n)$?

*5. Let $f(t, x)$ be an irreducible polynomial in two variables over $\mathbf{Q}$, and let $\mathcal{H}$ be the set of all rational numbers α such that $f(\alpha, x)$ is irreducible as a polynomial over $\mathbf{Q}$ in x. $\mathcal{H}$ is called a *basic Hilbert set*. Lemma 36d shows that $\mathcal{H}$ is infinite. Prove that $\mathcal{H}$ is dense in the real line $\mathbf{R}$. (A set $\mathcal{S} \subset \mathbf{R}$ is said to be *dense* in $\mathbf{R}$ if for every $r \in \mathbf{R}$ there exists a sequence of numbers $s_n \in \mathcal{S}$ converging to r.)

Section 4.3. Existence of Polynomials over Q with Group S_n

In this section we shall see that for every positive integer n there exists a polynomial over $\mathbf{Q}$ whose Galois group is S_n. As noted earlier, this implies that for every $n \geqslant 5$, there is a polynomial over $\mathbf{Q}$ none of whose roots can be expressed in radicals. Another implication will be to show that for every $m \geqslant 2$, there exist numbers α such that $\deg_{\mathbf{Q}} \alpha = 2^m$ but α is not constructible. Although a number of the preliminary results we need have been anticipated by problems in Chapter 3, they will be repeated here so that the entire argument may be more easily comprehended.

Let **F** be a field. We say that the n complex numbers a_1, $a_2, \ldots, a_n$ are *algebraically independent over* **F** if there is no nontrivial polynomial $p(x_1, x_2, \ldots, x_n)$ in n variables over **F** such that $p(a_1, a_2, \ldots, a_n) = 0$. In other words, there is no algebraic relation among the numbers a_1, $a_2, \ldots, a_n$.

LEMMA 37a. *If* **F** *is a countable field, then for every n there exist n algebraically independent elements over* **F**.

Proof. Since **F** is countable, the set of numbers algebraic over **F** is also countable (Problem 4, Section 1.5). Since **C** is uncountable, we can find a number a_1 which is transcendental over **F**. The extension $\mathbf{F}(a_1)$ is simply the set $\{f(a_1)/g(a_1) \mid f, g \in \mathbf{F}[x], g \neq 0\}$, and it is an elementary countability exercise to show that this set is countable. Thus we can apply the same argument to find an element a_2 which is transcendental over $\mathbf{F}(a_1)$. Continuing in this fashion we can generate a sequence of numbers $a_1, a_2, \ldots, a_n$ such that a_n is transcendental over $\mathbf{F}(a_1, a_2, \ldots, a_{n-1})$. To show that these numbers are algebraically independent, let us suppose to the contrary that there is a nontrivial polynomial relation among them: $p(a_1, a_2, \ldots, a_n) = 0$. If k is the largest subscript such that a_k actually appears in this relation, then we could conclude that a_k is algebraic over $\mathbf{F}(a_1, a_2, \ldots, a_{k-1})$, a contradiction. ∎

Next we see how permutations of algebraically independent elements give rise to field automorphisms.

LEMMA 37b. *Let* $a_1, a_2, \ldots, a_n$ *be algebraically independent over a field* **F**, *and let* $\mathbf{E} = \mathbf{F}(a_1, a_2, \ldots, a_n)$. *Then* *each of the $n!$ permutations of the list $a_1, a_2, \ldots, a_n$ induces an automorphism on* **E** *which leaves* **F** *fixed.*

Proof. Any permutation can be accomplished by a sequence of transpositions (Problem 8, Section 1.6). Thus it suffices to show that every transposition induces an automorphism, for the composition of automorphisms is an automorphism. Moreover, it is apparent that the sequence of simple extensions leading from **F** to **E** by each of the elements $a_1, a_2, \ldots, a_n$ may be taken in any order. Therefore, in treating the transposition which interchanges a_i and a_j, we may for convenience define **K** as the extension of **F** by all the other a's, and then write $\mathbf{E} = \mathbf{K}(a_i, a_j) = \mathbf{K}(c, d)$, where we have set $c = a_i$, $d = a_j$.

The set **E** may be represented as $\mathbf{E} = \{p(c, d)/q(c, d) \mid p, q$ polynomials over **K** in two variables, $q \neq 0\}$. The natural candidate for the automorphism ϕ induced by the transposition which interchanges c and d is the mapping

$$\phi: \frac{p(c, d)}{q(c, d)} \to \frac{p(d, c)}{q(d, c)} .$$

It is necessary to show that this mapping is well defined, one-to-one and onto, that it preserves addition and multiplication, and that it leaves elements of **F** unchanged.

To show that ϕ is well defined, we must consider the possibility that an element of **E** has two different representations, say

$$\frac{p_1(c, d)}{q_1(c, d)} = \frac{p_2(c, d)}{q_2(c, d)} .$$

We want to show in this case that both representations

lead to the same value of the image under ϕ, that is,

$$\frac{p_1(d, c)}{q_1(d, c)} = \frac{p_2(d, c)}{q_2(d, c)} .$$

The first equation implies that (c, d) is a solution to the polynomial equation $p_1q_2 - p_2q_1 = 0$. But since c and d are algebraically independent, the polynomial $p_1q_2 - p_2q_1$ must be identically 0. Therefore (d, c) is also a solution, from which we deduce the second equation.

By reversing the argument we also see that ϕ is one-to-one. It is obvious that ϕ is onto and that it leaves **F** fixed. To show that ϕ preserves addition, we need to verify that

$$\phi\left[\frac{p_1(c, d)}{q_1(c, d)} + \frac{p_2(c, d)}{q_2(c, d)}\right] = \frac{p_1(d, c)}{q_1(d, c)} + \frac{p_2(d, c)}{q_2(d, c)} .$$

To apply ϕ on the left side, we first need to cross multiply and add the fractions; then we interchange c and d, after which we can rewrite again as a sum. It should be obvious that this yields the expression on the right, and so a formal argument is omitted. Preservation of multiplication is verified in the same way. ∎

From now on we shall take the field **F** to be **Q**. Since **Q** is countable, for any n we can choose n algebraically independent elements over **Q**. The symbols $a_1, a_2, \ldots, a_n$ will be used throughout to represent a specific choice of such elements, and we shall let $\mathbf{E} = \mathbf{Q}(a_1, a_2, \ldots, a_n)$. For each i from 1 to n, let us define

$$b_i = (-1)^i \sigma_i(a_1, a_2, \ldots, a_n),$$

where σ_i is the ith elementary symmetric function of n

variables (Section 1.6). Thus the numbers $a_1, a_2, \ldots, a_n$ are the roots of the polynomial $f(x) = x^n + b_1x^{n-1} + \cdots + b_n$, which is a polynomial over the field $\mathbf{K} = \mathbf{F}(b_1, b_2, \ldots, b_n)$.

LEMMA 37c. *The polynomial f is irreducible over* $\mathbf{K}$. *Its group over* $\mathbf{K}$, *namely* $G(\mathbf{E}/\mathbf{K})$, *is* S_n.

Proof. The first statement follows immediately from the second. By Lemma 37b, each permutation of the a_i's gives rise to an element of $G(\mathbf{E}/\mathbf{F})$. However, the fixed field is actually larger. Since each permutation leaves symmetric functions invariant, all the b_i's are left fixed, and consequently so too are all the elements of $\mathbf{K}$. ∎

Since $\mathbf{E}$ is a finite extension of $\mathbf{K}$, it can be written as a simple extension. The next lemma asserts that this can be done in a particularly simple way:

LEMMA 37d. *There exist integers $m_1, m_2, \ldots, m_n$ such that* $\mathbf{E} = \mathbf{K}(m_1a_1 + m_2a_2 + \cdots + m_na_n)$. *The sum $m_1a_1 + m_2a_2 + \cdots + m_na_n$ assumes $n!$ different values under the $n!$ possible permutations of the a_i's.*

Proof. In the proof of Theorem 18, it was shown how multiple extensions may be written as simple extensions. The argument there actually shows that for all but a finite number of n-tuples $(m_1, m_2, \ldots, m_n)$, the above simple extension equals all of $\mathbf{E}$.

Since $[\mathbf{E} : \mathbf{K}] = |G(\mathbf{E}/\mathbf{K})| = n!$, the degree of $\sum_{i=1}^n m_ia_i$ over $\mathbf{K}$ is $n!$. Thus it has $n!$ distinct conjugates. However, the conjugates are the images of $\sum_{i=1}^n m_ia_i$ under the automorphisms in $G(\mathbf{E}/\mathbf{K})$, and such automorphisms map

$\sum_{i=1}^{n} m_i a_i$ to similar sums with the a_i's permuted. Since the conjugates must be distinct, all these sums must be distinct. ∎

Let us denote by $c_1, c_2, \ldots, c_{n!}$ the $n!$ distinct values mentioned in the previous lemma. We define $g(x) = \prod_{i=1}^{n!}(x - c_i)$.

LEMMA 37e. *The polynomial g is irreducible over* **K**. *Its group over* **K** *is* S_n.

Proof. Since permutations of a_i's only lead to permutations of the c_i's, the coefficients of g are symmetric polynomials evaluated at the a_i's. Since the a_i's are the roots of a polynomial over **K**, these coefficients are in **K** (Corollary 9a). The group of the polynomial g over **K** is also $G(\mathbf{E}/\mathbf{K}) = S_n$. From this it follows that g is irreducible. ∎

Thus we note that both f and g, of respective degrees n and $n!$, have group S_n over **K**. We are now going to construct two polynomials $F(t_1, t_2, \ldots, t_n, x)$ and $G(t_1, t_2, \ldots, t_n, x)$ in $n + 1$ variables over **Q**. They will have the property that their respective degrees in x are also n and $n!$ and that substitution of the values $t_i = a_i$ for each i reduces them to the polynomials f and g.

In particular, let us begin with n variables $s_1, s_2, \ldots, s_n$. We define n functions t_i by the equation

$$t_i = (-1)^i \sigma_i(s_1, s_2, \ldots, s_n)$$

for each i from 1 to n. We also define $n!$ functions u_i by the sum $\sum_{i=1}^{n} m_i s_i$ and all the analogous sums obtained by permuting the s_i's. Using these variables, we can construct F and G:

$$F(t_1, t_2, \ldots, t_n, x) = x^n + t_1 x^{n-1} + t_2 x^{n-2} + \cdots + t_n$$

$$= \prod_{i=1}^{n} (x - s_i),$$

$$G(t_1, t_2, \ldots, t_n, x) = \prod_{i=1}^{n!} (x - u_i).$$

Whereas we have a simple explicit representation of F in terms of $t_1, t_2, \ldots, t_n$ and x, for G we do not. We know that the coefficients of the powers of x in G may be written as polynomials in the s_i's, and that they are symmetric in the s_i's. By the Fundamental Theorem on Symmetric Functions, then, we do know that G is a function of $t_1, t_2, \ldots, t_n$, and x. The polynomial F is called the *generic polynomial of nth degree.*

LEMMA 37f. *The polynomials F and G are irreducible over* **Q**.

Proof. Suppose we could write a nontrivial factorization

$$F(t_1, t_2, \ldots, t_n, x)$$

$$= R(t_1, t_2, \ldots, t_n, x) S(t_1, t_2, \ldots, t_n, x)$$

where R and S also have rational coefficients. Then this formula would have to hold for all particular values of t_1 through t_n, and specifically when for each i, $t_i = b_i$, with b_i as defined earlier. If R and S each have degree at least one in the variable x, then the substitution $t_i = b_i$ gives a factorization of f, a contradiction to Lemma 37c. If S has degree 0 in x, say, then the coefficient of x^n in the product has the form $r(t_1, \ldots, t_n)s(t_1, \ldots, t_n)$, and this must equal 1. Since the degree of rs is the sum of the degrees of r and s, and since this sum must be 0, we conclude that r and s are nonzero constants. But if s is a constant, so is S; and thus the factorization $F = RS$ is a trivial one.

The argument for G is identical. ∎

We shall actually only need the part of this lemma which concerns G.

It is now rather simple to give the answer to our question about polynomials over **Q** with group S_n.

THEOREM 37. *For every positive integer n there exists a polynomial over* **Q** *whose Galois group is* S_n.

Proof. By Hilbert's Irreducibility Theorem, we may choose rational numbers $\beta_1, \beta_2, \ldots, \beta_n$ such that the polynomial $\tilde{G}(x) = G(\beta_1, \beta_2, \ldots, \beta_n, x)$ is irreducible in $\mathbf{Q}[x]$. We shall see that the polynomial $\tilde{F}(x) = F(\beta_1, \beta_2, \ldots, \beta_n, x) = x^n + \beta_1 x^{n-1} + \beta_2 x^{n-2} + \cdots + \beta_n$ has Galois group S_n.

Let $\alpha_1, \alpha_2, \ldots, \alpha_n$ denote the roots of $\tilde{F}$. Then the number $m_1\alpha_1 + m_2\alpha_2 + \cdots + m_n\alpha_n$ is in the splitting field $\mathbf{Q}(\alpha_1, \alpha_2, \ldots, \alpha_n)$. By the definition of G, this number is one of the β_i's. Therefore it is a root of $\tilde{G}$, which is an irreducible polynomial of degree $n!$ over **Q**. Therefore $[\mathbf{Q}(\alpha_1, \alpha_2, \ldots, \alpha_n) : \mathbf{Q}] \geqslant n!$ and so the Galois group of $\tilde{F}$ has order $\geqslant n!$. But since this group corresponds to a set of permutations of $\alpha_1, \alpha_2, \ldots, \alpha_n$, its order is also at most $n!$. In order for the group to have this order, therefore, it must be precisely S_n. ∎

Let us state as corollaries the implications of this theorem for the problems of solvability by radicals and constructibility.

COROLLARY 37a. *For every* $n \geqslant 5$ *there exists a polynomial over* **Q** *of degree n such that none of its roots can be expressed in radicals.*

Proof. By Theorem 37 there always exists a polynomial over **Q** with group S_n. For $n \geqslant 5$, the group S_n is not solvable (Lemma 35b). By Galois' Theorem, the polynomial is not solvable by radicals, meaning that at least one root cannot be expressed in radicals. But since the group is S_n, the polynomial is irreducible (Problem 6, Section 3.6). Therefore (Problem 5, Section 3.6), none of the roots can be expressed in radicals. ∎

Recall that in Theorem 20 it was shown that a necessary condition for a number to be constructible is that its degree over **Q** be a power of 2. The following corollary shows that this condition is not sufficient.

COROLLARY 37b. *For every* $m \geqslant 2$, *there exists a number which has degree* 2^m *over* **Q** *but which is not constructible.*

Proof. Let $n = 2^m$ and let $f(x)$ be a polynomial over **Q** with group S_n. Since f is irreducible, each of its roots has degree $n = 2^m$ over **Q**. If all these roots were constructible, then every element of the splitting field **E** of f would be constructible, for all such elements can be expressed as rational combinations of the roots of f. The degree $[\mathbf{E} : \mathbf{Q}]$ is $n!$, however, since it equals the order of the Galois group. If we write **E** as a simple extension of **Q**, $\mathbf{E} = \mathbf{Q}(\alpha)$, then $\deg_{\mathbf{Q}}\alpha = n!$. But since $n \geqslant 4$, $n!$ contains the odd factor 3, and thus $n!$ is not a power of 2. By Theorem 20 then, α is not constructible. Thus at least one of the roots of f is not constructible. ∎

An example of such a number has already been encountered in Problem 5 of Section 3.1.

PROBLEMS

1. In the context of Lemma 37b, are there any automorphisms of **E** leaving **F** fixed other than those induced by the $n!$ permutations of the a_i's?
2. Are the elements $b_1, b_2, \ldots, b_n$, as defined in the text, algebraically independent over **Q**?
3. Let $H(t_1, t_2, \ldots, t_n, x)$ be a polynomial over **Q** in $n+1$ variables. Let $b_1, b_2, \ldots, b_n$ be n complex numbers which are algebraically independent over **Q**. Define a polynomial $h(x) = H(b_1, b_2, \ldots, b_n, x)$, and consider it over $\mathbf{K} = \mathbf{Q}(b_1, b_2, \ldots, b_n)$. What is the relation between the irreducibility of H and the irreducibility of h?
4. Show that none of the roots of the polynomial determined in the proof of Corollary 37b can be constructed.

REFERENCES AND NOTES

Basic material on power series may be found in advanced calculus texts, such as [1, 2], for the real case, and in complex variables texts, such as [3, 4], for the complex case. References for our small excursion into algebraic geometry are [5, 6]. Hilbert's original paper on the irreducibility theorem is [7]. Our treatment follows [8, 9] except that by working at regular points it has been possible to avoid introducing

Puiseux expansions with fractional exponents. An example of a polynomial over $\mathbf{Q}$ with group S_6 is $f(x) = x^6 + 22x^5 - 9x^4 + 12x^3 - 37x^2 - 29x - 5$, for which see p. 109 of [10]. Generalizations of the basic problem of this chapter are discussed on p. 262 of [11].

1. Angus E. Taylor and W. Robert Mann, *Advanced Calculus*, 2nd ed., Xerox, Lexington, Mass., 1972.

2. John M. H. Olmsted, *Real Variables*, Appleton-Century-Crofts, New York, 1959.

3. Ruel V. Churchill and James W. Brown, *Complex Variables and Applications*, rev. ed., McGraw-Hill, New York, 1974.

4. Norman Levinson and Raymond M. Redheffer, *Complex Variables*, Holden-Day, San Francisco, 1970.

5. Martin Eichler, *Introduction to the Theory of Algebraic Numbers and Functions*, tr. by G. Striker, Academic Press, New York, 1966.

6. A. Seidenberg, *Elements of the Theory of Algebraic Curves*, Addison-Wesley, Reading, Mass., 1968.

7. David Hilbert, Ueber die Irreducibilität ganzer rationaler Functionen mit ganzzahligen Coefficienten, *J. Reine Angew. Math.*, 110 (1892) 104–129.

8. Karl Dörge, Einfacher Beweis des Hilbertschen Irreduzibilitätssatzes, *Math. Ann.*, 96 (1927) 176–182.

9. Serge Lang, *Diophantine Geometry*, Interscience, New York, 1962.

10. Nathan Jacobson, *Lectures in Abstract Algebra*, vol. III, Van Nostrand, Princeton, N. J., 1964.

11. ——, *Basic Algebra I*, Freeman, San Francisco, 1974.

SOLUTIONS TO THE PROBLEMS

Section 1.1

1. *Use Lemma 1a to show that if a and b are constructible, then so too are $a + b$, $a - b$, ab, and, when $b \neq 0$, a/b.*

By hypothesis, segments of lengths $|a|$ and $|b|$ may be constructed, beginning with a unit segment. To obtain the constructibility of the number $a + b$, we want to construct a segment of length $|a + b|$. If a and b have the same sign, $|a + b| = |a| + |b|$, in which case the addition part of Lemma 1a gives the result. If a and b have opposite signs, then either $|a + b| = |a| - |b|$ or $|a + b| = |b| - |a|$, depending on which difference is positive. In this case the subtraction portion of Lemma 1a applies. That $a - b$ is constructible follows at once from the preceding, since $a - b = a + (-b)$ and if b is constructible, then by the definition of constructibility so is $-b$. Multiplication is immediate from Lemma 1a, since $|ab| = |a| \cdot |b|$; and so too is division.

2. *If M and N are positive integers, show that if $\sqrt[M]{N}$ is rational, then, in fact, it must be an integer. Conclude then that $\sqrt{2}$, $\sqrt{3}$, $\sqrt[3]{2}$, $\sqrt{6}$, for example, are all irrational. (Hint: If $\sqrt[M]{N}$ is rational it may be expressed in lowest terms as Q/R. If $R \neq \pm 1$ it has a prime factor. If a prime divides a product of integers, it must divide one of them.)*

Following the hint, if $\sqrt[M]{N} = Q/R$ in lowest terms (say with R positive), then $NR^M = Q^M$. If a prime p divides R, then it divides NR^M and hence Q^M. But then it must be a factor of Q. This contradicts the fact that Q/R is in lowest terms. Hence $R = 1$.

By inspection, there are no integers equal to $\sqrt{2}$, $\sqrt{3}$, $\sqrt[3]{2}$, $\sqrt{6}$. Hence they are irrational.

*3. *We know from Lemma* 1c *that if* **F** *is a field, so too is the set* $\{a + b\sqrt{2} \mid a, b \in \mathbf{F}\}$. *What about the set* $\{a + b\sqrt[3]{2} \mid a, b \in \mathbf{F}\}$? *Is it ever a field? Is it always a field? Give as complete an analysis as you can.*

It is a field if and only if $\sqrt[3]{2} \in \mathbf{F}$. For, if $\sqrt[3]{2} \in \mathbf{F}$, it is just **F** itself. If it is a field, then by closure under multiplication, $(\sqrt[3]{2})^2$ must also be an element of the field. Hence there must exist a and b in **F** for which $2^{2/3} = a + b2^{1/3}$. Therefore $(2^{1/3})^2 - b(2^{1/3}) - a = 0$, and so $2^{1/3} = (b \pm \sqrt{b^2 + 4a})/2$. Now, if $\sqrt{b^2 + 4a} \in \mathbf{F}$, it follows that $\sqrt[3]{2} \in \mathbf{F}$. If $\sqrt{b^2 + 4a} \notin \mathbf{F}$, then $\sqrt[3]{2}$ is of the form $A + B\sqrt{k}$ with A, B, and k in **F**, $\sqrt{k}$ not in **F**. That is, $2 = (A + B\sqrt{k})^3 = C + D\sqrt{k}$ for some C and D in **F**. But $D = 0$, for if not, then $\sqrt{k} = (2 - C)/D$ which is in **F**, a contradiction. But if we compute $(A - B\sqrt{k})^3$ we get

$$\begin{aligned}(A - B\sqrt{k})^3 &= C - D\sqrt{k} \\ &= C \\ &= 2,\end{aligned}$$

so that $(A + B\sqrt{k})$ and $(A - B\sqrt{k})$ are both real cube roots of 2. By elementary calculus, there is a unique real cube root of 2, and so $A + B\sqrt{k} = A - B\sqrt{k}$. Therefore $B = 0$, for if it were not, we would have $\sqrt{k} = 0 \in \mathbf{F}$, a contradiction. Thus $\sqrt[3]{2} = A \in \mathbf{F}$, which completes the proof. Hence for $\mathbf{F} = \mathbf{Q}$, it is not a field, by the irrationality of $\sqrt[3]{2}$. For $\mathbf{F} = \mathbf{R}$, it is a field, namely $\mathbf{R}$ itself.

4. *Let $\mathfrak{F}$ be a collection of fields. Show that the common intersection of all the fields in $\mathfrak{F}$, written $\bigcap_{\mathbf{F} \in \mathfrak{F}} \mathbf{F}$, is itself a field.*

Call the intersection S. If a and b are in S, then for each $\mathbf{F} \in \mathfrak{F}$ they are in $\mathbf{F}$. Since $\mathbf{F}$ is a field, $a + b, a - b, ab$, and, when $b \neq 0$, a/b, are also in $\mathbf{F}$. Since this holds for every $\mathbf{F}$, they are also in S. Since each $\mathbf{F}$ also contains 1, so too does S. Hence S is a field.

*5. *Give an explicit description of the smallest field containing $\sqrt[3]{2}$. (By the "smallest" field having a given property is meant the intersection of all fields with that property. Problem 4 guarantees that such an intersection is a field.)*

The answer may be expressed as $\mathbf{F} = \{a + b2^{1/3} + c2^{2/3} \mid a, b, c \in \mathbf{Q}\}$. It should be clear that the desired field $\mathbf{F}$ must contain all elements of the form $a + b2^{1/3} + c2^{2/3}$. It remains then to show the set of such elements alone is itself a field. Closure under addition and subtraction is evident; it is almost as clear that the product of two such numbers has the same form, as 2 can be factored out of any term $2^{4/3}$ in a product. The difficulty is in showing closure under division. This will follow from closure under multiplication if we can just show whenever $a + b2^{1/3} + c2^{2/3} \neq 0$, which happens to be the case exactly when a, b, and c are not all 0, then there exist rationals A, B, and C such that

$$\frac{1}{a + b2^{1/3} + c2^{2/3}} = A + B2^{1/3} + C2^{2/3}.$$

In fact, by multiplying both sides of the above by a

suitable rational (as the reader should check), we can assume without loss of generality that a, b, and c are integers and are relatively prime (that is, they have no common factor other than ± 1). (Of course, we cannot expect that A, B, and C will also be of this form.) Multiplying the equation through by $a + b2^{1/3} + c2^{2/3}$ and grouping terms yields

$$1 = [aA + 2cB + 2bC] + [bA + aB + 2cC]2^{1/3}$$
$$+ [cA + bB + aC]2^{2/3},$$

which will have a rational solution A, B, C if the three linear equations

$$aA + 2cB + 2bC = 1$$
$$bA + aB + 2cC = 0$$
$$cA + bB + aC = 0$$

have a rational solution in the unknowns A, B, and C. This will certainly be the case if the determinant of the coefficients

$$\Delta = \begin{vmatrix} a & 2c & 2b \\ b & a & 2c \\ c & b & a \end{vmatrix}$$

is not 0, for then the solutions can be calculated from a, b, and c by rational operations, from which it follows that they are rational. We will show that the assumption $\Delta = 0$ leads to a contradiction of the fact that a, b, and c are relatively prime. Expanding Δ by cofactors of the first column, or by any other method, we obtain

$$\Delta = a(a^2 - 2bc) - b(2ac - 2b^2) + c(4c^2 - 2ab)$$
$$= a^3 - 6abc + 2b^3 + 4c^3.$$

Now, if $\Delta = 0$, then 2 divides a^3, as it clearly divides the other terms. Hence 2 divides a. But then 4 divides the first, second, and fourth terms, and so 2 divides b^3 and hence b. But then 8 divides each of the first three terms, and so 2 divides c^3 and hence c. Thus 2 is a common factor of a, b, and c. As noted earlier, this contradiction completes the proof.

*6. *If p is a prime and N is a positive integer $\geqslant 2$, give an explicit description of the smallest field containing $\sqrt[N]{p}$. (This is a generalization of the previous problem.)*

The answer is $F = \{\sum_{j=0}^{n-1} a_j p^{j/N} \mid \text{each } a_j \in \mathbf{Q}\}$. As in the previous problem, the desired field clearly must contain this set. Thus it remains to show that this set is itself a field. Closure under addition, subtraction, and multiplication is readily established; but the difficulty lies in establishing closure under division. This reduces to closure under multiplication if we can show that every nonzero element has a reciprocal of the same form. That is, we want to show that whenever $\sum_{j=0}^{N-1} a_j p^{j/N} \neq 0$ (which happens to be the case exactly when not all the a_j's are 0), then there exist rationals A_j, $0 \leqslant j \leqslant N-1$, such that

$$\frac{1}{a_0 + a_1 p^{1/N} + a_2 p^{2/N} + \cdots + a_{N-1} p^{(N-1)/N}}$$
$$= A_0 + A_1 p^{1/N} + A_2 p^{2/N} + \cdots + A_{N-1} p^{(N-1)/N}.$$

By multiplying both sides by a suitable rational (in fact the reciprocal of the least common denominator of all the a_j's, when expressed as the quotients of integers) we can reduce the problem to the case in which the a_j's are all integers. Then, by multiplying both sides by the greatest common divisor of the a_j's, the problem is reduced to the case where the a_j's are relatively prime (that is, they have no common factor other than ± 1). Thus, without loss of generality, we shall establish the existence of appropriate A_j's for the case where the a_j's are relatively prime integers. Multiplying through by the denominator above, and grouping appropriately, we obtain the equation

$$\begin{aligned}1 = &\left[a_0A_0 + pa_{N-1}A_1 + pa_{N-2}A_2 + \cdots + pa_1A_{N-1}\right]\\ &+ \left[a_1A_0 + a_0A_1 + pa_{N-1}A_2 + \cdots + pa_2A_{N-1}\right]p^{1/N}\\ &+ \left[a_2A_0 + a_1A_1 + a_0A_2 + \cdots + pa_3A_{N-1}\right]p^{2/N}\\ &+\\ &\vdots\\ &+ \left[a_{N-1}A_0 + a_{N-2}A_1 + a_{N-3}A_2\right.\\ &\qquad \left.+ \cdots + a_0A_{N-1}\right]p^{(N-1)/N}.\end{aligned}$$

This will be solved by any solution of the following simultaneous linear equations in the unknowns $A_0, A_1, \ldots, A_{N-1}$:

$$\begin{aligned}a_0A_0 + pa_{N-1}A_1 + pa_{N-2}A_2 + \cdots + pa_1A_{N-1} &= 1\\ a_1A_0 + a_0A_1 + pa_{N-1}A_2 + \cdots + pa_2A_{N-1} &= 0\\ a_2A_0 + a_1A_1 + a_0A_2 + \cdots + pa_3A_{N-1} &= 0\\ &\vdots\\ a_{N-1}A_0 + a_{N-2}A_1 + a_{N-3}A_2 + \cdots + a_0A_{N-1} &= 0.\end{aligned}$$

A solution certainly exists if the determinant of the coefficients is not 0, in which case the values of the A_j's will all be rational, as follows from the fact that their explicit calculation only involves the application of rational operations to the coefficients. Thus our solution will be complete once it is shown that the determinant

$$\Delta = \begin{vmatrix} a_0 & pa_{N-1} & pa_{N-2} & \cdots & pa_1 \\ a_1 & a_0 & pa_{N-1} & \cdots & pa_2 \\ a_2 & a_1 & a_0 & \cdots & pa_3 \\ \vdots & & & & \\ a_{N-1} & a_{N-2} & a_{N-3} & \cdots & a_0 \end{vmatrix}$$

cannot be 0. This will be done by showing that the condition $\Delta = 0$ leads to a contradiction of the assumption that the a_j's are relatively prime.

The basic tool is the usual definition of a determinant as a sum of products, each prefixed by an appropriate sign (plus or minus) and containing as factors exactly one element from each row and each column of the array. Observe that each term of the expansion of Δ contains explicitly the factor p, except for the term a_0^N. Thus (under the assumption $\Delta = 0$), p divides a_0^N, and hence p divides a_0. Thus, in Δ, we may replace a_0 by pa_0' for some integer a_0':

$$\Delta = \begin{vmatrix} pa_0' & pa_{N-1} & pa_{N-2} & \cdots & pa_1 \\ a_1 & pa_0' & pa_{N-2} & \cdots & pa_2 \\ a_2 & a_1 & pa_0' & \cdots & pa_3 \\ \vdots & & & & \\ a_{N-1} & a_{N-2} & a_{N-3} & \cdots & pa_0' \end{vmatrix}.$$

But now every term of the expansion contains explicitly the factor p^2, except for the term pa_1^N, gotten by taking the last element of the first row and then all the elements just below the main diagonal. Thus p divides a_1^N, and so p divides a_1. Continuing in this fashion, it is easy to see that

p divides each a_j, which is a contradiction to their being relatively prime.

7. *Is* $\{a + b\sqrt{2} + c\sqrt{3} \mid a, b, c \in \mathbf{Q}\}$ *a field*?

No, because it would have to contain $\sqrt{6}$ (by closure under multiplication), and this is impossible. To see this, suppose that for some rational numbers a, b, and c, we could write $\sqrt{6} = a + b\sqrt{2} + c\sqrt{3}$. It would follow that

$$\sqrt{6} - a = b\sqrt{2} + c\sqrt{3},$$

$$(\sqrt{6} - a)^2 = (b\sqrt{2} + c\sqrt{3})^2,$$

$$6 - 2a\sqrt{6} + a^2 = 2b^2 + 2bc\sqrt{6} + 3c^2,$$

$$6 + a^2 - 2b^2 - 3c^2 = 2(a + bc)\sqrt{6}.$$

This contradicts the irrationality of $\sqrt{6}$ except possibly when $a + bc = 0$. But in this case, we also have

$$6 + a^2 - 2b^2 - 3c^2 = 0,$$

$$6 + (-bc)^2 - 2b^2 - 3c^2 = 0,$$

$$b^2(c^2 - 2) = 3(c^2 - 2),$$

and so either $c^2 = 2$ or $b^2 = 3$, which are both impossible by the irrationality of $\sqrt{2}$ and $\sqrt{3}$.

8. *Give an explicit representation of the smallest field containing both* $\sqrt{2}$ *and* $\sqrt{3}$. (*Hint*: *Use one of the lemmas.*)

The required field $\mathbf{F}$ is

$$\{a + b\sqrt{2} + c\sqrt{3} + d\sqrt{6} \mid a, b, c, d \in \mathbf{Q}\}.$$

To see this, we develop **F** by a sequence of extensions. Let $\mathbf{F}_1 = \mathbf{Q}(\sqrt{2})$ and $\mathbf{F} = \mathbf{F}_1(\sqrt{3})$. By the closure property, **F**, which by Lemma 1c is a field, must be the answer. For, any field containing $\sqrt{2}$ and $\sqrt{3}$ must contain all the elements of **F**. As to the representation of the elements of **F**, it consists of the set of elements of the form $(a + b\sqrt{2}) + (c + d\sqrt{2})\sqrt{3}$, where a, b, c, and $d \in \mathbf{Q}$, and thus is given precisely by the original description.

9. *For the example treated after Lemma* 1*d*, *give a different development of the same number from the rationals which results in a different sequence of field extensions.*

The development:

$$13, \sqrt{13}, 6, \sqrt{6}, 7, \sqrt{7}, \sqrt{\sqrt{13}} = \sqrt[4]{13}, \tfrac{4}{3},$$

$$1, 2, 2\sqrt{7}, 1 + 2\sqrt{7}, \sqrt{1 + 2\sqrt{7}}, \sqrt{6} + \sqrt{1 + 2\sqrt{7}},$$

$$\tfrac{4}{3}, \sqrt{\sqrt{6} + \sqrt{1 + 2\sqrt{7}}}, \tfrac{4}{3}\sqrt{\sqrt{6} + \sqrt{1 + 2\sqrt{7}}},$$

$$\sqrt[4]{13} + \tfrac{4}{3}\sqrt{\sqrt{6} + \sqrt{1 + 2\sqrt{7}}}$$

gives the sequence of extensions:

$$\mathbf{F}_0 = \mathbf{Q}, \mathbf{F}_1 = \mathbf{F}_0(\sqrt{13}), \mathbf{F}_2 = \mathbf{F}_1(\sqrt{6}),$$

$$\mathbf{F}_3 = \mathbf{F}_2(\sqrt{7}), \mathbf{F}_4 = \mathbf{F}_3\left(\sqrt{\sqrt{13}}\right),$$

$$\mathbf{F}_5 = \mathbf{F}_4\left(\sqrt{1 + 2\sqrt{7}}\right), \mathbf{F}_6 = \mathbf{F}_5\left(\sqrt{\sqrt{6} + \sqrt{1 + 2\sqrt{7}}}\right).$$

10. *Let* **F** *be a field. Is the set* $\{a + b\sqrt{k} \mid a, b, k \in \mathbf{F}, k > 0\}$ *necessarily a field*? (*This differs from* $\mathbf{F}(\sqrt{k})$

because in the present case k is not fixed, but rather it ranges through all positive values in $\mathbf{F}$.

It is not a field, since it is not necessarily even closed under addition. For suppose $\mathbf{F} = \mathbf{Q}$. Then $\sqrt{2} + \sqrt{3}$ does not belong to this set, even though both $\sqrt{2}$ and $\sqrt{3}$ individually do. To see this, suppose we could write $\sqrt{2} + \sqrt{3} = a + b\sqrt{k}$ for some rationals a, b, and k, $k > 0$. Without loss of generality, we may assume k is an integer, as we can always replace $\sqrt{M/N}$ by $(1/N)\sqrt{M \cdot N}$. Squaring once, we obtain

$$2 + 2\sqrt{6} + 3 = a^2 + 2ab\sqrt{k} + b^2k.$$

If either a or b is 0, this contradicts the irrationality of $\sqrt{6}$. Otherwise we have

$$2\sqrt{6} - 2ab\sqrt{k} = a^2 + b^2k - 5$$

$$\left(2\sqrt{6} - 2ab\sqrt{k}\right)^2 = \left(a^2 + b^2k - 5\right)^2$$

$$24 - 8ab\sqrt{6k} + 4a^2b^2k = \left(a^2 + b^2k - 5\right)^2$$

from which it follows that $\sqrt{6k}$ is rational and hence, by Problem 2, an integer. By considering the prime power decomposition of k, we see that 2 and 3, and hence 6, divide k; from which it follows that $k = 6M^2$ for some integer M. Thus we have

$$5 + 2\sqrt{6} = a^2 + 2abM\sqrt{6} + b^2(6M^2),$$

or

$$5 - 6M^2 6^2 - a^2 = (2abM - 2)\sqrt{6}.$$

So, by the irrationality of $\sqrt{6}$, the right and hence the left side are both 0. From $2abM - 2 = 0$, we get $M = 1/ab$.

Using this,

$$5 - 6M^2b^2 - a^2 = 0$$

$$5 - 6/a^2 - a^2 = 0$$

$$(a^2)^2 - 5a^2 + 6 = 0$$

$$a^2 = \frac{5 \pm \sqrt{25 - 24}}{2}$$

$$a^2 = 2 \text{ or } 3,$$

thus contradicting the irrationality of $\sqrt{2}$ or $\sqrt{3}$.

11. *The following argument is occasionally found in proofs of Theorem* 1: "*Using only points in the plane of a field* **F**, *by a single fundamental construction it is only possible to construct points in the plane of some* $\mathbf{F}(\sqrt{k})$, $k \in \mathbf{F}$." *Criticize this argument or use it to give a simpler proof of Theorem* 1.

It is ambiguous at best, fallacious at worst. The difficulty is that on a single fundamental construction, it is possible to obtain points in the planes of several different quadratic extensions of F, for the new circle or line may intersect any number of previous ones, each case possibly leading to a different extension.

12. *What else would you like to know? Spend a little time thinking about the material in this section, and, perhaps especially, in the above problems. Can you anticipate the solution of any of the classical construction problems: doubling the cube, trisecting an angle, squaring the circle? Are you curious about anything? Do any questions occur to you that seem to be interesting? Make a short list of your questions and ideas, spending at least enough time on each to satisfy yourself that it is nontrivial. Keep these ideas in*

mind, jotting down others as they occur to you as you proceed through the material in subsequent sections.

There are many possibilities. Three of the most striking may be the following. What is the answer to Problem 6 when p is not prime? For example, what about $N = 3$ and $p = 4$ or 6? What is the smallest field containing π? In connection with Problem 9, do all such sequences of extensions have the same number of fields? The observation that one can always toss in irrelevant steps requires that this question needs rephrasing to rule out such cases. Assuming this is done, now what is the answer? Incidentally, Problem 3 essentially solves the problem of doubling the cube.

Section 1.2

1. *Can the cube be "tripled"?*

No. This would require that $\sqrt[3]{3}$ be constructible. But, by the identical argument as that used to develop Lemma 2 and Theorem 2, this would imply that $\sqrt[3]{3}$ is rational, which is a contradiction.

2. *Can the cube be "quadrupled"?*

No. This would require that $\sqrt[3]{4} = (\sqrt[3]{2})^2$ be constructible. By Lemma 1b, then, $\sqrt[3]{2}$ would be constructible, which was shown to be impossible in Theorem 2.

Section 1.3

1. *It is obvious that by repeated bisection it is possible to divide an arbitrary angle into 4 equal parts. Show how this may also be deduced algebraically from the equation relating* $\cos 4\theta$ *and* $\cos \theta$.

We find $\cos 4\theta = 8\cos^4\theta - 8\cos^2\theta + 1$ and so $\cos^2\theta$

$= [8 \pm \sqrt{64 - 32(1 - \cos 4\theta)}\,]/16$. Since the operations involved here all correspond to possible constructions, $\cos^2\theta$ can be constructed if we are given $\cos 4\theta$. But then, by Lemma 1b, $\cos\theta$ can also be constructed.

2. *Discuss the validity of the following proof of Theorem* 3: "*By the continuity of* $\cos\theta$, *there exists an angle* θ_0 *such that* $\cos\theta_0 = (\sqrt[3]{2})/2$. *By Theorem* 2, $\cos\theta_0$ *is not constructible. Hence* $3\theta_0$ *is an angle that cannot be trisected.*"

It is invalid. Even though $\cos\theta_0$ cannot be constructed from a unit segment, it is conceivable that it might be constructed with segments equal to both 1 and $\cos 3\theta_0$ being given.

*3. *Prove that it is not possible, in general, to quintisect an arbitrary angle, i.e., to divide it into five equal parts.*

The angle between 0 and π whose cosine is $-5/6$ is constructible, since $-5/6$ is a constructible number. Call this angle 5θ. If 5θ were also quintisectible, then θ would be constructible, or equivalently $\cos\theta$ would be a constructible number. Setting $x = 2\cos\theta$, the trigonometric addition formulas yield $2\cos 5\theta = x^5 - 5x^3 + 5x$, so that in this case at least one of the roots of $f(x) = 3x^5 - 15x^3 + 15x + 5 = 0$ would be constructible. This polynomial has one root between 1.7 and 1.8 by the Intermediate Value Theorem, since $f(1.7) < 0$ and $f(1.8) > 0$ by computation. Call this root $2\cos\theta_1$. Thus $\cos\theta_1 > .85$ and from tables $0 < \theta_1 < 32°$. Letting

$$\theta_2 = \frac{2\pi}{5} - \theta_1, \qquad \theta_3 = \frac{2\pi}{5} + \theta_1,$$

$$\theta_4 = \frac{4\pi}{5} - \theta_1, \qquad \theta_5 = \frac{4\pi}{5} + \theta_1,$$

we make two observations. First, for all i, $\cos 5\theta_i = -5/6$, and so all the numbers $\cos\theta_i$ are roots of f. Second, the numbers θ_i are all on the interval from 0 to π and they are distinct (follows since $\theta_1 < \pi/5 = 36°$), and therefore the numbers $2\cos\theta_i$ are the complete set of five roots of f. Third, if any θ_i is constructible, they all are; this is because they differ by a multiple of $\pi/5$, which is a constructible angle ($\pi/5 = 36°$, one-half the central angle of a regular pentagon, which is well known to be constructible). Putting it all together, if 5θ is quintisectible, then all the roots of f are constructible numbers.

The rest of the solution makes use of two very simple facts called to the reader's attention in subsequent sections. The first is that the sum of the roots of a polynomial $a_n x^n + a_{n-1}x^{n-1} + \cdots + a_0$ is just $-a_{n-1}/a_n$. (See Section 1.6.) Thus the sum of the roots of f must be 0. The second is that if M/N is a rational root in lowest terms, then N must divide a_n and M must divide a_0. (See Problem 2, Section 1.5.) From this latter result we see that f has no rational roots, for the candidates are ± 1, $\pm 1/3$, ± 5, $\pm 5/3$, and none of these gives 0 when substituted into f.

Let N_1 be minimal such that there is a root of f in a field $\mathbf{F}_{N_1}$ obtainable from $\mathbf{Q}$ by a sequence of quadratic extensions $\mathbf{Q} = \mathbf{F}_0 \subset \mathbf{F}_1 \subset \mathbf{F}_2 \subset \cdots \subset \mathbf{F}_{N_1}$. Call such a root r_1; it has the form $a + b\sqrt{k}$, where $a, b, k \in \mathbf{F}_{N_1-1}$, $\sqrt{k} \notin \mathbf{F}_{N_1-1}$, and $b \neq 0$. If we compute $f(a + b\sqrt{k})$ we see that it has the form $A + B\sqrt{k}$, for some $A, B \in \mathbf{F}_{N_1-1}$. Since $A + B\sqrt{k} = 0$ and $\sqrt{k} \notin \mathbf{F}_{N_1-1}$, it follows that $B = 0$. Because f contains only odd powers of x, we find that $f(a - b\sqrt{k}) = A - B\sqrt{k} = A = A + B\sqrt{k} = 0$ also, so that $a - b\sqrt{k}$ is a second root of f. Let us call it r_2.

Suppose a third root r_3 of f also belongs to $\mathbf{F}_{N_1}$. By the same argument as above so too would a fourth root r_4.

But then $r_5 = -r_1 - r_2 - r_3 - r_4$ would be in $\mathbf{F}_{N_1-1}$, since the $\sqrt{k}$'s cancel on addition. This would contradict the minimality of N_1.

Thus there must be a minimal $N_2 > N_1$ such that we can continue our sequence of field extensions $\mathbf{F}_0 \subset \mathbf{F}_1 \subset \cdots \subset \mathbf{F}_{N_1} \subset \mathbf{F}_{N_1+1} \subset \cdots \subset \mathbf{F}_{N_2}$ until another root $r_3 \in \mathbf{F}_{N_2}$. As above $\mathbf{F}_{N_2}$ must contain also another root r_4. But then the fifth root, $r_5 = -r_1 - r_2 - r_3 - r_4$ would be in $\mathbf{F}_{N_2-1}$, contradicting the minimality of N_2.

Therefore none of the roots of f can be constructible, and this completes the proof.

4. *Is the difficulty in Problem 2 real or imagined? Show that there are trisectible angles 3θ such that θ is not constructible.*

We know $\cos 3\theta = 4\cos^3\theta - 3\cos\theta$. If we pick a θ_0 satisfying $\cos\theta_0 = (\sqrt[3]{2})/2$, then $\cos 3\theta_0 = 1 - (3/2)\sqrt[3]{2}$. By Theorem 2, θ_0 is not constructible; but it can obviously be constructed given segments equal to both 1 and $\cos 3\theta_0$.

Section 1.4

1. *The number $\sqrt{2}+\sqrt{3}$, being constructible, has associated with it a sequence of quadratic field extensions $\mathbf{Q} = \mathbf{F}_0 \subset \mathbf{F}_1 \subset \cdots \subset \mathbf{F}_N$ such that $\sqrt{2}+\sqrt{3} \in \mathbf{F}_N$. Determine the smallest such number N. Using this value, what is the degree of the polynomial whose existence is proved in Theorem 4? Following the procedure in the proof of Theorem 4, determine this and any intermediate polynomials.*

By the solution to Problem 10, Section 1.1, $N \geqslant 2$. That $N = 2$ will do is clear from taking $\mathbf{F}_1 = \mathbf{Q}(\sqrt{2})$, $\mathbf{F}_2 = \mathbf{F}_1(\sqrt{3})$. The proof of Theorem 4 gives a polynomial of degree $2^2 = 4$. It is determined as follows:

$$x - \sqrt{2} - \sqrt{3} = 0$$

$$x - \sqrt{2} = \sqrt{3}$$

$$x^2 - (2\sqrt{2})x + 2 = 3$$

which is the polynomial of degree 2 with coefficients in $\mathbf{F}_1$. Repeating the process,

$$x^2 - 1 = (2\sqrt{2})x$$

$$x^4 - 2x^2 + 1 = 8x^2$$

$$x^4 - 10x^2 + 1 = 0.$$

2. *Does there exist a polynomial equation with rational coefficients and of degree less than that determined in Problem* 1 *such that* $\sqrt{2} + \sqrt{3}$ *is a root*?

No. Suppose there were a lower degree polynomial equation with rational coefficients, such that $\sqrt{2} + \sqrt{3}$ is a root. Then for some rationals a, b, c, and d (not all 0) we would have:

$$0 = a(\sqrt{2} + \sqrt{3})^3 + b(\sqrt{2} + \sqrt{3})^2 + c(\sqrt{2} + \sqrt{3}) + d$$

$$= a[2\sqrt{2} + 6\sqrt{3} + 9\sqrt{2} + 3\sqrt{3}] + b[2 + 2\sqrt{6} + 3]$$

$$+ c[\sqrt{2} + \sqrt{3}] + d$$

$$= \sqrt{2}\,[11a + c] + \sqrt{3}\,[9a + c] + \sqrt{6}\,[2b] + [5b + d].$$

But by the solution to Problem 7, Section 1.1, $b = 0$. Similar calculations imply that $a = c = d = 0$. Thus there is no such equation.

Section 1.5

1. a. *If f is a polynomial of degree $n \geqslant 1$ over a field* **F**, *and if r is any element of* **F**, *show that there is a polynomial q of degree $n-1$ over* **F** *and a number R in* **F** *such that $f(x) = (x - r)q(x) + R$.*

b. *Show that, in the above, $R = f(r)$.*

c. *Let f be a polynomial over* **C**. *Show that r is a root of f if and only if $x - r$ is a factor of f.*

a. If $f(x) = a_n x^n + a_{n-1}x^{n-1} + \cdots + a_0$, we seek to find the coefficients of a polynomial $q(x) = A_{n-1}x^{n-1} + \cdots + A_0$ such that $f(x) = (x - r)q(x) + R$, for some number $R \in \mathbf{F}$. Thus, by equating coefficients of like powers of x, we obtain the equations: $a_n = A_{n-1}$, $a_{n-1} = -rA_{n-1} + A_{n-2}, \ldots, a_{n-j} = -rA_{n-j} + A_{n-j-1}, \ldots, a_0 = -rA_0 + R$. These may be solved in order, yielding appropriate A_j's and R.

b. In $f(x) = (x - r)q(x) + R$, let x take the value r. This gives $R = f(r)$.

c. The "if" part is obvious and the "only if" part is a consequence of the above.

2. *If f is a polynomial with integral coefficients it is sometimes useful to know whether it has any rational roots. Suppose M/N is a rational root expressed in lowest terms. There is a simple necessary condition connecting M and N with the first and last coefficients of the polynomial. Find it. (This result is called the Rational Roots Theorem. It gives a necessary condition that can be used to narrow down the candidates for a rational root to a small collection which can each be tested.)*

Writing $f(x) = a_n x^n + a_{n-1}x^{n-1} + \cdots + a_0$, if M/N is a rational root in lowest terms, then N must divide a_n and M must divide a_0. To see this, substitute M/N into the equation $f(x) = 0$, multiply through by N^n, and collect terms appropriately, thereby obtaining

$$a_n M^n = N[-a_{n-1}M^{n-1} - a_{n-2}M^{n-2}N - \cdots - a_0 N^{n-1}],$$

and

$$a_0 N^n = M[-a_n M^{n-1} - a_{n-1}M^{n-2}N - \cdots - a_1 N^{n-1}].$$

From the first, we see that N divides $a_n M^n$, but since N has no factors at all in common with M, it must divide a_n. (Just take the prime power decomposition of N and observe that the factors, one by one, each divide a_n.) From the second, by similar reasoning, we see that M must divide a_0.

3. *Use the results of Problem* 2 *to find any rational roots of each of the following*:

a. $8x^3 - 6x - 1$
b. $x^3 - 3x - 1$
c. $2x^3 + 3x^2 - x - 1$
d. $3x^5 - 5x^3 + 5x - 1$
e. $x^3 + 4x^2 + x - 6$.

a. Candidates: $\pm 1/8$, $\pm 1/4$, $\pm 1/2$, ± 1. None of these turns out to be a root.

b. Candidates: ± 1. Neither is a root. Notice how, as in Section 1.3, this and the previous equation are closely related, the roots of the second being twice the roots of the first. But the second is much easier to check for rational roots.

c. Candidates: $\pm 1/2$, ± 1. $-1/2$ is the only rational root.

d. Candidates: $\pm 1/3$, ± 1. None is a root.

e. Candidates: ± 6, ± 3, ± 2, ± 1. The roots are -3, -2, $+1$.

4. *If* **F** *is a countable field, show that the field of all numbers algebraic over* **F** *is also countable.*

There are many ways to do this; the reader may find the following method instructive. Let us first take the case where $\mathbf{F} = \mathbf{Q}$. Each algebraic number is the root of a polynomial with integral coefficients. So first we show that the set of such polynomials is countable. Denoting by $p_1, p_2, p_3, \ldots$ the sequence of primes (of which there are an infinite number), we associate with each nonzero polynomial $a_n x^n + a_{n-1} x^{n-1} + \cdots + a_0$ the rational number $p_1^{a_0} p_2^{a_1} p_3^{a_2} \cdots p_{n+1}^{a_n}$, which is in lowest terms. This is a one-to-one correspondence with the positive rationals excluding 1. Since such rationals form a countable set, the polynomials with integral coefficients can be written in a list: $f_1, f_2, f_3, \ldots$. To make a list of the algebraic numbers, we list first the roots of f_1, then of f_2, etc., ignoring repetitions. If **F** is any countable field, the polynomials over **F** can be put in an obvious one-to-one correspondence with those with integral coefficients. Thus we can also write a list $f_1, f_2, \ldots,$ of such polynomials, from which we see that the set of all their roots is countable.

5. *Show that the result of Problem* 2 *of Section* 1.1 *is just a special case of Problem* 2 *of this section.*

$\sqrt[M]{N}$ must be a root of $x^M - N = 0$. By the result of Problem 2 of this section, any rational root of this equation, when expressed in lowest terms, must have a denominator dividing 1, the leading coefficient. Hence the root must actually be an integer.

6. *If r is a root of multiplicity $\geqslant 2$ of a polynomial f, show that $f'(r) = 0$, where f' denotes the derivative of f.*

By the definition of multiplicity, we may write $f(x) = (x - r)^2 g(x)$. By the product rule for differentiation, $f'(x) = (x - r)^2 g'(x) + 2(x - r)g(x)$, from which it is clear that $f'(r) = 0$.

Section 1.6

1. *The cubic $2x^3 - 3x^2 - 32x - 15$ has two roots whose sum is 2. Find all the roots.*

The sum of all three roots is $-a_{n-1}/a_n = 3/2$, so one root is $-1/2$. Thus $(x + \frac{1}{2})$ is a factor, and by division we obtain $2x^3 - 3x^2 - 32x - 15 = (x + \frac{1}{2})(2x^2 - 4x - 30)$. Thus the other roots are the roots of the quadratic factor, which are found to be 5 and -3.

2. *Find the product of all the roots of $2x^5 - 3x^4 + 7x^3 - 7x^2 + 3x - 1$. Are any of the roots rational?*

The product of the roots is $(-1)^n a_0/a_n = 1/2$. By Problem 2, Section 1.5, the only candidates for rational roots are ± 1 and $\pm 1/2$. Calling the polynomial f, we find $f(1) = 1, f(-1) = -23, f(1/2) = -1/2$, and $f(-1/2) = -43/8$. So there are no rational roots.

3. *Find the value, expressed in terms of the coefficients, of the sum of the squares of thc roots of $a_n x^n + a_{n-1}x^{n-1} + \cdots + a_0$.*

Since the symmetric polynomial $x_1^2 + x_2^2 + \cdots + x_n^2$ may be expressed in terms of the elementary symmetric functions as $\sigma_1^2 - 2\sigma_2$, we get for the sum of the squares of the roots $(-a_{n-1}/a_n)^2 - 2a_{n-2}/a_n$, or $[a_{n-1}^2 - 2a_{n-2}a_n]/a_n^2$.

4. *Find the value, expressed in terms of the coefficients, of the sum of the reciprocals of the roots of* $a_n x^n + a_{n-1}x^{n-1} + \cdots + a_0$, *where* $a_0 \neq 0$.

The reciprocals of the roots of the given polynomial are the roots of the polynomial $a_0 x^n + a_1 x^{n-1} + \cdots + a_n$, the sum of whose roots is $-a_1/a_0$.

5. *Let* P_1 *and* P_2 *be two polynomials in n variables. Show that the highest term of the product* P_1P_2 *is the product of the highest terms of* P_1 *and* P_2. *(The word "highest" refers to the ordering on the exponent n-tuples.)*

This is an immediate consequence of the fact that inequalities involving exponent n-tuples can be "added" in the usual way. That is, suppose $(i_1, i_2, \ldots, i_n) \geqslant (i'_1, i'_2, \ldots, i'_n)$ and $(j_1, j_2, \ldots, j_n) \geqslant (j'_1, j'_2, \ldots, j'_n)$; then, from the definition of the ordering and the corresponding property for inequalities involving real numbers, $(i_1 + j_1, i_2 + j_2, \ldots, i_n + j_n) \geqslant (i'_1 + j'_1, i'_2 + j'_2, \ldots, i'_n + j'_n)$.

6. (*Alternate proof of Theorem* 8.) *Suppose a and b are algebraic over a field* **F**. *Then a is a root of some polynomial over* **F**, *whose complete set of roots may be written* $a = a_1, a_2, \ldots, a_n$, *where each root is listed as many times as its multiplicity. Similarly for b, giving rise to roots* $b = b_1, b_2, \ldots, b_m$. *Apply the theory of symmetric functions twice successively to the expression*

$$\prod_{i=1}^{n} \prod_{j=1}^{m} (x - a_i - b_j)$$

to conclude that $a + b$ *is algebraic over* **F**. *In a similar way, show that ab is also algebraic over* **F**.

Applying Corollary 9a to each coefficient in the polynomial $\prod_{j=1}^{m}(u - b_j)$, we see that it is a polynomial over **F**, call it f. Then $\prod_{i=1}^{n}\prod_{j=1}^{m}(x - a_i - b_j) = \prod_{i=1}^{n} f(x - a_i)$. Applying Corollary 9a to each coefficient in the expansion of this as a polynomial in x, we see that it is a polynomial over **F**. But $a + b$ is one of its roots. The product case is similar. Write

$$\prod_{i=1}^{n}\prod_{j=1}^{m}(x - a_i b_j) = \prod_{i=1}^{n} a_i^m \prod_{j=1}^{m}\left(\frac{x}{a_i} - b_j\right) = \prod_{i=1}^{n} a_i^m f(x/a_i),$$

where f is a polynomial over **F**. Since the degree of f is m, each coefficient of $a_i^m f(x/a_i)$ is an element of **F** multiplied by some nonnegative integral power of a_i. Taking the product over all i, there results a polynomial in x to each of whose coefficients Corollary 9a is applicable.

7. *Prove that every permutation of the variables* $x_1, x_2, \ldots, x_n$ *can be accomplished by a succession of transpositions, where a transposition is the single interchange of two of the variables. (This was used in the proof of Corollary 9b.)*

By induction on the number of variables. If $n = 1$, there is nothing to prove. Assuming it is true for $n - 1$ variables, proceed as follows. We want to transform the ordered n-tuple $(x_1, x_2, \ldots, x_n)$ into $(x_{\phi(1)}, x_{\phi(2)}, \ldots, x_{\phi(n)})$, where ϕ is a permutation on the first n integers, by a sequence of transpositions. For $j = \phi(n)$, interchange x_j and x_n. It remains to properly permute the remaining $n - 1$ variables, which can be done by a succession of transpositions, according to the inductive hypothesis.

8. *With reference to the preceding problem, there may be many different sequences of transpositions resulting in the same permutation. Show that for a given permutation, all sequences of transpositions leading to it have the same*

parity; that is, either they all contain an even number of transpositions or else they all contain an odd number. (In this way, permutations can be classified as "even" or "odd".)

Consider the behavior of the *alternating function* P of n variables, $P(x_1, x_2, \ldots, x_n) = \prod_{i<j}(x_i - x_j)$, under two sequences S and S' of transpositions leading to the same permutation. Writing P out,

$$\begin{aligned} P = {} & (x_1 - x_2)(x_1 - x_3)(x_1 - x_4) \cdots (x_1 - x_n) \\ & \cdot (x_2 - x_3)(x_2 - x_4) \cdots (x_2 - x_n) \\ & \cdots \\ & \cdot (x_{n-1} - x_n), \end{aligned}$$

we see that any transposition merely changes the sign of P. For, suppose x_i and x_j are interchanged, with $i < j$. Then the factors involving neither x_i nor x_j are left unaltered; the factor $(x_i - x_j)$ is replaced by $(x_j - x_i) = -(x_i - x_j)$; and the remaining factors may be paired off to form products of the form $\pm(x_i - x_k)(x_j - x_k)$, $k \neq i$ or j, each of which is unaltered by the transposition. Now suppose that S contains N transpositions and S' contains N' transpositions. Under the resulting permutation, P is transformed to $(-1)^N P$ and $(-1)^{N'} P$. Since these must be the same, N and N' must both be odd or both be even.

Section 1.7

1. *Prove that* $e^{z_1}e^{z_2} = e^{z_1+z_2}$.

Letting $z_1 = x_1 + iy_1$ and $z_2 = x_2 + iy_2$, we want to show that

$$\begin{aligned} & e^{x_1}(\cos y_1 + i \sin y_1)e^{x_2}(\cos y_2 + i \sin y_2) \\ & \qquad = e^{x_1+x_2}(\cos(y_1 + y_2) + i \sin(y_1 + y_2)). \end{aligned}$$

Since x_1 and x_2 are real, $e^{x_1}e^{x_2} = e^{x_1+x_2}$, so it suffices to show that $(\cos y_1 + i \sin y_1)(\cos y_2 + i \sin y_2) = \cos(y_1 + y_2) + i \sin(y_1 + y_2)$. Multiplying out on the left and using the fact that $i^2 = -1$, this simply reduces to the usual sine and cosine addition formulas.

2. *Prove that the binomial coefficients are in fact integers.*

For $N \geqslant 1$ and $0 \leqslant K \leqslant N$, we want to show that the binomial coefficient $\binom{N}{K}$, given by $\binom{N}{K} = N!/K!(N-K)!$, is an integer. By a simple calculation, for $1 \leqslant K \leqslant N-1$,

$$\binom{N}{K} = \binom{N-1}{K-1} + \binom{N-1}{K}.$$

For $K = 0$ or N, $\binom{N}{K} = 1$. The result now follows at once by induction on N.

3. *The modulus of a complex number $z = x + iy$, where x and y are real, is given by $|z| = \sqrt{x^2 + y^2}$. If $F(z)$ is a complex function of a complex variable, then the definition of the statement $\lim_{z \to z_0} F(z) = L$ is that for every $\epsilon > 0$ there exists a $\delta > 0$ such that $|F(z) - L| < \epsilon$ whenever $0 < |z - z_0| < \delta$. This is formally identical with the real case, the difference being the use of moduli instead of absolute values. If $G(z)$ is a complex function of a complex variable, the derivative of G is the new function whose value at any point z_0 is given by*

$$G'(z_0) = \lim_{z \to z_0} \frac{G(z) - G(z_0)}{z - z_0}.$$

(As in the real case, some complex functions are differentiable and some are not.) Show that the usual rules from elementary calculus carry over in the complex case for each of the following:

a. *the derivatives of sums, differences, products, and quotients*;
b. *the derivatives of polynomials*;
c. *the derivatives of e^z and e^{-z}.*

a. The calculations are identical with those from elementary calculus, so we simply work out the details for one part, say the product rule. Given two differentiable functions u and v, we want to show that $(uv)'(z_0) = u(z_0)v'(z_0) + v(z_0)u'(z_0)$. We have:

$$(uv)'(z_0) = \lim_{z \to z_0} \frac{u(z)v(z) - u(z_0)v(z_0)}{z - z_0}$$

$$= \lim_{z \to z_0} \frac{u(z)v(z) - u(z)v(z_0) + u(z)v(z_0) - u(z_0)v(z_0)}{z - z_0}$$

$$= \lim_{z \to z_0} u(z)\left[\lim_{z \to z_0} \frac{v(z) - v(z_0)}{z - z_0}\right]$$

$$+ v(z_0) \lim_{z \to z_0}\left[\frac{u(z) - u(z_0)}{z - z_0}\right]$$

$$= u(z_0)v'(z_0) + v(z_0)u'(z_0).$$

We have of course used the basic properties of limits as well as the continuity of u at z_0, all of which are very straightforward.

b. The results of the previous part, together with the observation that the derivative of a constant is 0, reduce the problem to showing that for $n \geqslant 1$, the derivative of z^n is nz^{n-1}. For $n = 1$ it is obvious. The induction step follows from the product rule: $(z^n)' = (z \cdot z^{n-1})' = z \cdot (n-1)z^{n-2} + 1 \cdot z^{n-1} = nz^{n-1}$.

c. From the definition of the derivative and from Problem 1,

$$(e^z)'(z_0) = \lim_{z \to z_0} \frac{e^z - e^{z_0}}{z - z_0} = e^{z_0} \lim_{h \to 0} \frac{e^h - 1}{h},$$

where $h = z - z_0$ is complex. It remains to show that this limit equals 1. Writing $h = a + ib$, with $a, b \in \mathbf{R}$, this limit becomes

$$\lim_{(a,b)\to(0,0)} \frac{e^a(\cos b + i \sin b) - 1}{a + ib} \cdot \frac{(a - ib)}{(a - ib)}$$

$$= \lim_{(a,b)\to(0,0)} \left[\frac{ae^a \cos b - a + be^a \sin b}{a^2 + b^2} \right]$$

$$+ i \lim_{(a,b)\to(0,0)} \left[\frac{b - be^a \cos b + ae^a \sin b}{a^2 + b^2} \right].$$

We want to show that the first limit equals 1 and the second equals 0. These results follow after appropriate algebraic manipulation, coupled with the observation that $|2ab| \leqslant a^2 + b^2$ and the following elementary real limits:

$$\lim_{b \to 0} \frac{1 - \cos b}{b} = 0, \ \lim_{b \to 0} \frac{\sin b}{b} = 1, \text{ and } \lim_{a \to 0} \frac{e^a - 1}{a} = 1.$$

The derivative of e^{-z} can now be obtained by the quotient rule, since $e^{-z} = 1/e^z$.

4. *Let $\phi(u)$ be a complex-valued function of the real variable u on the interval $[a, b]$ and suppose its real and imaginary parts are given by differentiable functions $r(u)$ and $s(u)$, respectively, so that*

$$\phi(u) = r(u) + is(u).$$

Then the derivative $\phi'(u)$ is defined by the equation

$$\phi'(u) = r'(u) + is'(u).$$

Similarly, the integral $\int_a^b \phi'(u)du$ is defined by the equation

$$\int_a^b \phi'(u)du = \int_a^b r'(u)du + i\int_a^b s'(u)du.$$

Prove the relevant form of the Fundamental Theorem of Calculus:

$$\phi(b) - \phi(a) = \int_a^b \phi'(u)du.$$

From elementary calculus, $\int_a^b r'(u)du = r(b) - r(a)$, and similarly for s. Thus $\int_a^b \phi'(u)du = r(b) - r(a) + is(b) - is(a) = \phi(b) - \phi(a)$.

5. *Let x be a fixed complex number, g a differentiable complex valued function of a complex variable, and u a real variable. Then the function $\phi(u) = g(ux)$ is a complex function of a real variable. Show that $\phi'(u) = g'(ux)\cdot x$.*

The definition of $\phi'(u)$ from Problem 4 clearly implies that for each u_0,

$$\phi'(u_0) = \lim_{u\to u_0} \frac{\phi(u) - \phi(u_0)}{u - u_0},$$

from which we deduce

$$\phi'(u_0) = \lim_{u\to u_0}\left[\frac{g(ux) - g(u_0x)}{ux - u_0x}\cdot x\right]$$

$$= g'(u_0x)\cdot x.$$

6. *In which part(s) of the proof of Theorem* 10 *was it necessary to use the fact that* p *is prime?*

Only in the proof of Claim 3.

7. *It was implicitly assumed in the proof of Theorem* 10 *that there are an infinite number of primes. Where was this assumed? Prove this fact.*

It was assumed in the final step, yielding the contradiction. That is, it was assumed that there actually exists a prime larger than k, a, and a_0 and such that $CB^{p-1}/(p-1)!$ can be made arbitrarily small.

Suppose that there were only a finite collection of primes, $p_1, p_2, \ldots, p_N$. Then the number $1 + \prod_{i=1}^{N} p_i$ would have to be prime, not being divisible by any of $p_1, p_2, \ldots, p_N$. But it is too large to equal any of the primes in the list, a contradiction.

8. *Prove that the number* e *is transcendental. (Hint: This is quite a bit easier than the corresponding proof for* π, *but can be developed along the same outline. In particular, the assumption that* e *is algebraic leads immediately to an equation very similar to the equation* $e^{q_1} + e^{q_2} + \cdots + e^{q_m} + k = 0$, *but where the exponents are integers and there may be integral coefficients in front of the terms. Working backwards, the analogue of* g *can be constructed, and from this* f *and* F. *The proof proceeds almost as before, but without the complication of symmetric functions, except in summing, some coefficients need to be inserted. This problem should help the reader gain a better understanding of the proof for* π.

If e were algebraic, it would satisfy an equation of the

form

$$b_m e^m + b_{m-1} e^{m-1} + \cdots + b_1 e^1 + k = 0,$$

where k and the b_j's are integers, $k > 0$. Define $g(x) = \prod_{j=1}^m (x - j)$. Defining f and F as in the proof for π, so that $a = 1$, we have

$$F(x) - e^x F(0) = -\int_0^1 x e^{(1-u)x} f(ux)\,du.$$

For each j, substitute in $x = j$ and multiply by b_j; summing the results we obtain

$$\sum_{j=1}^m b_j F(j) + kF(0) = -\sum_{j=1}^m \int_0^1 j b_j e^{(1-u)j} f(uj)\,du.$$

The term on the right is treated exactly as in the proof for π, in order to show that it can be made arbitrarily small for p a sufficiently large prime. The term $kF(0)$ is also treated as earlier, it not being divisible by p for any prime greater than or equal to both k and $m!$. Thus, it suffices to show that $\sum_{j=1}^m b_j F(j)$ is divisible by p, and here the proof is much simpler than in the case of π. In particular,

$$\sum_{j=1}^m b_j F(j) = \sum_{j=1}^m b_j \sum_{r=0}^{s+p} f^{(r)}(j) = \sum_{j=1}^m b_j \sum_{r=p}^{s+p} f^{(r)}(j),$$

since lower order derivatives of f have a factor g, which is 0 at each j. But for $r \geqslant p$, each coefficient of $(p-1)! \cdot f^{(r)}(x)$ is divisible by $p!$, and so each coefficient of $f^{(r)}(x)$ is divisible by p. Plugging in integral values of x, namely $x = j$, $1 \leqslant j \leqslant m$, yields for the sum an integer divisible by p. As noted earlier, this completes the argument.

Section 2.1

1. *Prove the division algorithm for* $\mathbf{Z}$: *If* $m, n \in \mathbf{Z}$, $n \neq 0$, *then there exist unique integers* q *and* r, *with* $0 \leqslant r < |n|$, *such that* $m = nq + r$.

If $n > 0$, the value $q = [m/n]$, where the brackets denote the greatest integer $\leqslant m/n$, certainly works, for then $r = m - nq$ is within the required interval. To show this, we first observe that $m - nq = m - n[m/n] \geqslant m - n(m/n) = 0$. But also, $m - nq = m - n[m/n] < m - n((m/n) - 1) = n$, since $[m/n] > (m/n) - 1$. To show uniqueness of q and r, suppose that $m = nq_1 + r_1$ and that $m = nq_2 + r_2$, with both r_1 and r_2 within the required bounds. Then $n(q_1 - q_2) = r_2 - r_1$, so that if $q_1 \neq q_2$, r_2 and r_1 are at least n units apart, which is impossible since they are both within the interval from 0 to $n - 1$. Hence $q_1 = q_2$, from which it follows that $r_1 = r_2$.

Now, for $n < 0$, by the above there are unique integers $\tilde{q}$ and r such that $m = (-n)\tilde{q} + r$, with $0 \leqslant r < |n|$. Simply take $q = -\tilde{q}$.

2. *Prove the Euclidean algorithm for* $\mathbf{Z}$: *If* $m, n \in \mathbf{Z}$, $n \neq 0$, *then there exist integers* s *and* t *such that the g.c.d.* $(m, n) = sm + tn$. *Moreover, the g.c.d. can be obtained in a finite number of steps by the method of repeated division of the most recent divisor by the most recent remainder, beginning with the division of* m *by* n *according to the division algorithm, as outlined in the text just prior to Theorem* 13.

By the division algorithm, the repeated division is certainly always possible. Since each remainder is at least one unit closer to 0 than the previous one, the process terminates (i.e., gives a 0 remainder) after at most n

divisions. Thus we have:

$$m = nq_1 + r_1$$

$$q_1 = r_1 q_2 + r_2$$

$$r_1 = r_2 q_3 + r_3$$

$$\vdots$$

$$r_{N-2} = r_{N-1} q_N + r_N$$

$$r_{N-1} = r_N q_{N+1} + 0.$$

Exactly as in the text, r_N can be written in the form $sm + tn$, from which we see that (m, n) divides r_N. But, beginning with the last equation and working backwards, we see that r_N divides $r_{N-1}, r_{N-2}, \ldots, q_1, n$, and m. Hence r_N divides (m, n). Thus we conclude that $r_N = (m, n)$.

3. *Find the g.c.d. of* 264 *and* 714. *Can you take them in either order*?

Yes, you can take them in either order, as the Euclidean algorithm makes no assumption about their relative size. Let us verify this here by doing the computations beginning both ways: $264 = 714 \cdot 0 + 264$, $714 = 264 \cdot 2 + 186$, $264 = 186 \cdot 1 + 78$, $186 = 78 \cdot 2 + 30$, $78 = 30 \cdot 2 + 18$, $30 = 18 \cdot 1 + 12$, $18 = 12 \cdot 1 + 6$, $12 = 6 \cdot 2 + 0$. If we begin with $714 = 264 \cdot 2 + 186$, we see that this is just the second step in the above sequence. We find that the g.c.d. is 6.

4. *Use the Euclidean algorithm for* $\mathbf{Q}[x]$ *to find a g.c.d. of* $f(x) = 2x^6 - 10x^5 + 2x^4 - 7x^3 - 15x^2 + 3x - 15$ *and* $g(x) = x^5 - 4x^4 - 3x^3 - 9x^2 - 4x - 5$.

By ordinary long division of polynomials, we find that $f = g\cdot(2x - 2) + (5x^3 - 25x^2 + 5x - 25)$, $g = (5x^3 - 25x^2 + 5x - 25)\cdot(x^2/5 + x/5 + 1/5)$, so that $(f, g) = 5x^3 - 25x^2 + 5x - 25$.

5. *Show that the g.c.d. of two polynomials is independent of the field over which we consider them to be. That is, suppose* $\mathbf{F}_1$ *and* $\mathbf{F}_2$ *are fields and* f *and* g *are polynomials belonging to both* $\mathbf{F}_1[x]$ *and* $\mathbf{F}_2[x]$; *show that they have the same g.c.d. in both* $\mathbf{F}_1[x]$ *and* $\mathbf{F}_2[x]$. (*Note: Some divisibility properties, such as irreducibility, are closely related to the field in question, but the g.c.d. is seen here not to be.*)

Let $\mathbf{F}_3 = \mathbf{F}_1 \cap \mathbf{F}_2$. The application of the Euclidean algorithm in either $\mathbf{F}_1[x]$ or $\mathbf{F}_2[x]$ is equivalent to its application in $\mathbf{F}_3[x]$, so the same result is gotten in both cases.

6. *Let* $f, g \in \mathbf{F}[x]$, f *irreducible. Suppose there exists some* $a \in \mathbf{C}$ *such that* $f(a) = g(a) = 0$; *that is, they have a common root in* $\mathbf{C}$. *Show that* f *divides* g (*in* $\mathbf{F}[x]$).

Their g.c.d. is nonconstant, since in $\mathbf{C}[x]$ it is divisible by $(x - a)$. Since f is irreducible, their g.c.d. must be f itself. Hence $f \mid g$.

7. *Let* $f, g, h \in \mathbf{F}[x]$, $f \neq 0$ (*i.e., for emphasis,* f *is not the* 0 *polynomial, although it may be* 0 *for certain values of* x). *If* $fg = fh$, *show that* $g = h$.

By the division algorithm, the quotient in dividing fg or fh by f is unique. Hence $g = h$.

8. *Show that* $(x^n - 1)/(x - 1)$ *is a polynomial over* $\mathbf{Q}$ *for all* (*positive integers*) n, *and determine precisely the set of values of* n *for which it is irreducible.*

Since 1 is a root of $(x^n - 1)$, $(x - 1)$ divides $(x^n - 1)$ in $\mathbf{C}[x]$ by Problem 1, Section 1.4. But since $(x - 1)$ and $(x^n - 1)$ are both in $\mathbf{Q}[x]$, the quotient will also be in $\mathbf{Q}[x]$. In fact, we easily see that $(x^n - 1)/(x - 1) = x^{n-1} + x^{n-2} + \cdots + x + 1$. If n is prime, this is irreducible, for the substitution $x = u + 1$ transforms it to the form

$$\begin{aligned}[(u+1)^n - 1]/u &= \left[u^n + \binom{n}{1}u^{n-1} + \binom{n}{2}u^{n-2} + \cdots + \binom{n}{n-1}u + 1 - 1\right]/u \\ &= u^{n-1} + \binom{n}{1}u^{n-2} + \binom{n}{2}u^{n-3} + \cdots + \binom{n}{n-1},\end{aligned}$$

in which each coefficient, except the leading one, is divisible by n. To see the latter, we write out

$$\binom{n}{k} = \frac{n!}{k!(n-k)!},$$

which is integral; since n is prime, $[k!(n-k)!] \mid (n-1)!$, so that n divides $\binom{n}{k}$. The last term is just n itself, so it is clearly not divisible by n^2. By the Eisenstein criterion, the polynomial in u, and hence that in x, is irreducible. If n is not prime, in which case we can write $n = jk$, we have

$$\begin{aligned}\frac{x^n - 1}{x - 1} &= \frac{x^n - 1}{x^k - 1} \cdot \frac{x^k - 1}{x - 1} \\ &= \left[(x^k)^{j-1} + (x^k)^{j-2} + \cdots + 1\right] \\ &\quad \times \left[x^{k-1} + x^{k-2} + \cdots + 1\right],\end{aligned}$$

which shows that for such n, the polynomial is reducible.

9. *Decompose* $x^{10}-1$ *into a product of irreducible factors in* $\mathbf{Q}[x]$.

First we use the fact that it is a difference of squares, and then we apply the result of the previous problem, thus obtaining

$$x^{10}-1=(x^5+1)(x^4+x^3+x^2+x+1)(x-1).$$

But since -1 is a root of x^5+1, $x+1$ must divide x^5+1, so we further obtain

$$x^{10}-1=(x^4-x^3+x^2-x+1)(x+1)\\ \times(x^4+x^3+x^2+x+1)(x-1).$$

The last three factors are already known to be irreducible. The irreducibility of the first factor may be established by the same method as in the previous problem, or from that result by the change of variables $v=-x$.

10. *Let* f *be a polynomial irreducible over* $\mathbf{F}$. *Show that* f *has no multiple roots. (Hint: Consider the g.c.d. of* f *and its derivative.)*

If a were a multiple root of f, then by the definition of multiplicity we could write $f=(x-a)^2g$, for some $g\in\mathbf{C}[x]$. Hence $f'(a)=0$, since $f'=(x-a)^2g'+2(x-a)g$. Hence, (f,f') has $(x-a)$ as a factor in $\mathbf{C}[x]$, and thus its degree is $\geqslant 1$. But the g.c.d. divides f in $\mathbf{F}[x]$. This is impossible, since its degree is also $\leqslant \deg f'<\deg f$, and f is irreducible.

11. *A polynomial with integral coefficients is said to be primitive if its coefficients are relatively prime (that is, there is no integer other than* ± 1 *which divides all the coefficients). If* f *is any polynomial with integral coefficients and if* g *is a primitive polynomial which divides* f, *show that*

the quotient f/g actually has integral coefficients. (As an application of this, observe that if a monic polynomial with integral coefficients divides another polynomial with integral coefficients, then the quotient has integral coefficients.)

This follows from the proof of Lemma 12. In particular, using the notation of that proof, the fact that g has integral coefficients implies that $B = 1$, and the fact that the coefficients are relatively prime implies that $A = 1$. Thus, since $E = \pm 1$, it follows that $f = gh = \pm g[(D/C)h]$, and so the polynomial $(D/C)h$, which has integral coefficients, must be $\pm h$ itself.

The application of this follows from the observation that a monic polynomial with integral coefficients is primitive.

12. *If f is a polynomial with integral coefficients and if g is a monic polynomial with integral coefficients, show that the quotient q and the remainder r, resulting from the division of f by g according to the division algorithm, both have integral coefficients.*

This additional property may be carried through in the induction argument used to prove the division algorithm. In particular, in that proof, $b_m = 1$, so that the reduced polynomial $f - (a_n/b_m)x^{n-m}g$ also has integral coefficients.

*13. *Develop an algorithm for factoring any polynomial in $\mathbf{Q}[x]$ into irreducible factors.*

Given a polynomial $f \in \mathbf{Q}[x]$, we give a procedure for finding a nontrivial factor, if it has one. By division, we can then write f as a product, and this essentially completes the problem, for the procedure can then be reapplied to each of the resulting factors. To begin, observe that multiplying through by an appropriate integer, we may transform f into a polynomial with integral coefficients; and since the factorization of this new polynomial is essentially equivalent to that of the original one, we assume without loss of generality that f already has integral coefficients. Let n denote the degree of f. We seek a polynomial $g(x) = \sum_{k=0}^{m} a_k x^k$ such that $g \mid f$, $a_m \neq 0$, and $1 \leqslant m \leqslant n-1$. By Gauss' Lemma, it suffices to consider the case where g and f/g have integral coefficients.

By trial and error, pick $n+1$ integers $x_0, x_1, \ldots, x_n$ such that none of them is a root of f. (This involves trying at most $2n+1$ integers, since f has at most n distinct roots.) Since g and f have integral coefficients, for every j, $g(x_j)$ and $f(x_j)$ are themselves integers. Since f/g has integral coefficients, $g(x_j) \mid f(x_j)$ in $\mathbf{Z}$. Since each $f(x_j)$ is a known integer, it has only a finite number of factors, which we are able to list. To put this last fact slightly differently, there are only a finite (and known) set of n-tuples $(i_0, i_1, \ldots, i_n)$ such that for every j, $i_j \mid f(x_j)$. Now, for each such n-tuple there is a unique polynomial g of degree $\leqslant n$ such that for every j, $g(x_j) = i_j$. To find it, write $g(x) = \sum_{k=0}^{n} a_k x^k$, temporarily allowing $m = n$, and solve the simultaneous equations $g(x_j) = i_j$, $0 \leqslant j \leqslant n$, for the coefficients, the a_k's. The determinant of the system is

$$\begin{vmatrix} 1 & x_0 & x_0^2 & \cdots & x_0^n \\ 1 & x_1 & x_1^2 & \cdots & x_1^n \\ \vdots & & & & \\ 1 & x_n & x_n^2 & \cdots & x_n^n \end{vmatrix} = \prod_{i>j} (x_i - x_j),$$

which is a Vandermonde determinant, and nonsingular since the x_j's are distinct. Thus there is a unique solution for the a_k's. If $a_n \neq 0$, or if only $a_0 \neq 0$, the resulting g is not a relevant candidate, since its degree would be n or 0; and in this case it would be discarded. In each other case, the resulting g may be divided into f according to the division algorithm to see whether the remainder is 0. Once a divisor of f is obtained, the procedure is repeated on the resulting factors of f. If f has a divisor, one must be obtained by this procedure. Thus, if no divisor is obtained, one can conclude that f is irreducible.

(This algorithm is primarily of theoretical interest; it would be very time-consuming to carry it out in most cases.)

Section 2.2

1. *If* **E** *is an extension of* **F**, *verify that* **E** *may be considered as a vector space over* **F**. (*A set V is said to be a vector space over a field* **F** *if it has the following properties*: (*a*) *there is a rule for adding vectors, always resulting in an element of V*; (*b*) *this addition operation is associative and commutative*; (*c*) *there is a vector called* 0, *such that $v+0=v$ for every v*; (*d*) *for every v there is a vector called $-v$, such that $v+(-v)=0$*; (*e*) *there is a rule, called scalar multiplication, for multiplying a scalar times a vector, always resulting in an element of V*; (*f*) *for $a, b \in \mathbf{F}$ and $u, v \in V$, $(ab)v = a(bv)$, $a(u+v) = au + av$, and $(a+b)v = av + bv$*; (*g*) *for every $v \in V$, $1v = v$, where* 1 *is simply the number* 1 *in* **F**.

Every one of the properties follows immediately from the definition of a field and the properties of complex addition and multiplication.

2. *Show that if* **E** *is a quadratic extension of* **F**, *then* $[\mathbf{E} : \mathbf{F}] = 2$.

By the definition of quadratic extension, $\mathbf{E} = \mathbf{F}(\sqrt{k})$, where $k \in \mathbf{F}$ but $\sqrt{k} \notin \mathbf{F}$. But $\sqrt{k}$ is a root of an irreducible polynomial of degree 2 over $\mathbf{F}$, namely $x^2 - k$. It is irreducible because if it factored over $\mathbf{F}$, the factors would be linear, and this is impossible since neither of its roots, $\pm\sqrt{k}$, are in $\mathbf{F}$. By Theorem 17, $[\mathbf{E} : \mathbf{F}] = 2$.

3. *Let* $\mathbf{E}$ *be an extension of* $\mathbf{F}$. *Show that* $[\mathbf{E} : \mathbf{F}] = 1$ *if and only if* $\mathbf{E} = \mathbf{F}$.

If $\mathbf{E} = \mathbf{F}$, then the single number 1 is a basis for $\mathbf{E}$ over $\mathbf{F}$, and so $[\mathbf{E} : \mathbf{F}] = 1$. If $\mathbf{E} \neq \mathbf{F}$, let a be some element in $\mathbf{E}$ but not in $\mathbf{F}$. Then the numbers a and 1 are linearly independent over $\mathbf{F}$, so that $[\mathbf{E} : \mathbf{F}] \geqslant 2$. (Of course, $[\mathbf{E} : \mathbf{F}]$ may be infinite.)

4. *If* $\mathbf{E}$ *is an extension of* $\mathbf{F}$ *such that every element of* $\mathbf{E}$ *is algebraic over* $\mathbf{F}$, *we say that* $\mathbf{E}$ *is algebraic over* $\mathbf{F}$. *Is every algebraic extension a finite extension*?

No; for example, let $\mathbf{E}$ be the set of all algebraic numbers and let $\mathbf{F} = \mathbf{Q}$. Then for any n, the n algebraic numbers $1, 2^{1/n}, 2^{2/n}, \ldots, 2^{(n-1)/n}$ are linearly independent, since $\deg 2^{1/n} = n$ by the irreducibility of $x^n - 2$ over $\mathbf{Q}$. Thus $\mathbf{E}$ contains linearly independent sets with arbitrarily many elements.

5. *If* $\mathbf{E}$ *is an algebraic extension of* $\mathbf{K}$, *and* $\mathbf{K}$ *is an algebraic extension of* $\mathbf{F}$, *show that* $\mathbf{E}$ *is an algebraic extension of* $\mathbf{F}$. (*See the previous problem for the definition of an algebraic extension.*)

Pick an arbitrary $a \in \mathbf{E}$. We want to show that a is algebraic over $\mathbf{F}$. Now, a is a root of some polynomial f over $\mathbf{K}$, since $\mathbf{E}$ is algebraic over $\mathbf{K}$. Let the coefficients of this polynomial be $a_n, a_{n-1}, \ldots, a_0$. Thus, a is actually algebraic over $\tilde{\mathbf{K}} \equiv \mathbf{F}(a_n, a_{n-1}, \ldots, a_0)$, which is a multiple algebraic extension of $\mathbf{F}$ and hence a finite extension of $\mathbf{F}$ By Theorem 19, $[\tilde{\mathbf{K}}(a) : \mathbf{F}] = [\tilde{\mathbf{K}}(a) : \tilde{\mathbf{K}}] \cdot [\tilde{\mathbf{K}} : \mathbf{F}]$, which is finite. Since a is in some finite extension of $\mathbf{F}$, by the first part of the proof of Theorem 18, a is algebraic over $\mathbf{F}$.

6. *Prove Theorem* 8 *by using the ideas of this section.* (*Hint*: *Observe that if a and b are algebraic over* $\mathbf{F}$, *then* $\mathbf{F}(a, b)$ *is a finite extension of* $\mathbf{F}$.)

The observation in the hint follows immediately from Theorem 18. Let c denote any of the numbers $a + b$, $a - b$, ab, or a/b (as long as $b \neq 0$). Since $c \in \mathbf{F}(a, b)$, we know that $\mathbf{F} \subset \mathbf{F}(c) \subset \mathbf{F}(a, b)$ and so $\deg_{\mathbf{F}} c = [\mathbf{F}(c) : \mathbf{F}]$ is finite by Theorem 19. By Theorem 17, c must then be algebraic over $\mathbf{F}$. Alternatively, apply to c the first observation in the proof of Theorem 18.

7. *Express* $\mathbf{Q}(\sqrt{2}, \sqrt{3})$ *as a simple extension of* $\mathbf{Q}$. *Find the degree of this extension over* $\mathbf{Q}$.

$\mathbf{Q}(\sqrt{2}, \sqrt{3}) = \mathbf{Q}(\sqrt{2} + \sqrt{3})$. It obviously suffices to show that $\sqrt{2}$ and $\sqrt{3}$ each belong to $\mathbf{Q}(\sqrt{2} + \sqrt{3})$. Define $a = \sqrt{2} + \sqrt{3}$. Then

$$a^2 = 5 + 2\sqrt{6}$$

$$\left(\frac{a^2 - 5}{2}\right)a = \sqrt{12} + \sqrt{18} = 2\sqrt{3} + 3\sqrt{2}$$

$$\left(\frac{a^2 - 5}{2}\right)a - 2a = \sqrt{2}$$

$$3a - \left(\frac{a^2 - 5}{2}\right)a = \sqrt{3}.$$

It was shown in Problems 1 and 2 of Section 1.4 that $\deg_{\mathbf{Q}}(\sqrt{2}+\sqrt{3})=4$, so this is the degree of the extension.

8. *Let f be a polynomial of degree n over* $\mathbf{F}$. *Show that there exists an extension* $\mathbf{E}$ *of* $\mathbf{F}$ *in which f is completely reducible (i.e., can be factored into linear factors) and such that* $[\mathbf{E}:\mathbf{F}] \leqslant n!$.

Let the roots of f (in $\mathbf{C}$) be $a_1, a_2, \ldots, a_n$. Define $\mathbf{F}_1=\mathbf{F}(a_1)$, $\mathbf{F}_2=\mathbf{F}_1(a_2), \ldots, \mathbf{F}_n=\mathbf{F}_{n-1}(a_n)$. $[\mathbf{F}_1:\mathbf{F}] \leqslant n$, since $\deg_{\mathbf{F}} a_1 \leqslant n$. Now factor $x-a_1$ out of f in $\mathbf{F}_1$, so that a_2 satisfies a polynomial of degree $n-1$ over $\mathbf{F}_1$. Thus $\deg_{\mathbf{F}_1} a_2=[\mathbf{F}_2:\mathbf{F}_1] \leqslant n-1$. Continue in this way and then apply Theorem 19, to obtain $[\mathbf{F}_n:\mathbf{F}]= [\mathbf{F}_n:\mathbf{F}_{n-1}]\cdot[\mathbf{F}_{n-1}:\mathbf{F}_{n-2}]\cdots[\mathbf{F}_1:\mathbf{F}] \leqslant (1)(2)\cdots(n-1)\cdot(n)=n!$

9. *Let a be transcendental over* $\mathbf{F}$. *Show that the set $\{a^k \mid k \in \mathbf{Z}\}$ is linearly independent in* $\mathbf{F}(a)$ *over* $\mathbf{F}$. *Is it a basis for* $\mathbf{F}(a)$ *over* $\mathbf{F}$?

To show linear independence, suppose we have a linear combination of elements a^k equal to 0. Multiplying through by the highest power of a in any denominator, we obtain a as a root of a polynomial over $\mathbf{F}$. Since a is transcendental over $\mathbf{F}$, it must be the 0 polynomial; thus the linear combination we started with must have all 0 coefficients.

This set is not a basis, since it does not span $\mathbf{F}(a)$. For example, $1/(a+1)$ cannot be written as a linear combination of the proposed elements. For if it could, there would be an $n>0$ such that

$$\frac{1}{a+1}=\sum_{k=-n}^{n} c_k a^k,\ c_k \in \mathbf{F}.$$

But then we could multiply both sides by $(a+1)a^n$ to get

a as the root of a nontrivial polynomial over $\mathbf{F}$, which would contradict its transcendence over $\mathbf{F}$.

10. *Let* $f \in \mathbf{R}[x]$ *with* $\deg f \geqslant 3$. *Show that* f *is reducible.*

Let a be any root of f. If f were irreducible, then $\deg_{\mathbf{R}} a = \deg f \geqslant 3$. However, $2 = [\mathbf{C} : \mathbf{R}] = [\mathbf{C} : \mathbf{R}(a)][\mathbf{R}(a) : \mathbf{R}]$, so that $[\mathbf{R}(a) : \mathbf{R}] = \deg_{\mathbf{R}} a = 1$ or 2. (An alternate proof would be to establish first that if a is a root of a polynomial over $\mathbf{R}$, then so too is $\bar{a}$, its complex conjugate. When f is completely factored in $\mathbf{C}[x]$, the factors corresponding to complex conjugate roots may be combined to give quadratic factors over $\mathbf{R}$. Thus the irreducible factors of f over $\mathbf{R}$ are linear and quadratic only. See the next problem.)

11. *Use the result of the previous problem to show that for* $f \in \mathbf{R}[x]$, *a is a root if and only if its complex conjugate $\bar{a}$ is a root. (The complex conjugate of* $b + ci$ *is* $b - ci$, *where* $b, c \in \mathbf{R}$.)

Since $\bar{\bar{a}} = a$, it suffices to show the "only if" part of the statement. If a is real, we are done, since then $\bar{a} = a$. If a is not real, then by Problem 10 it must be the root of a real quadratic factor of f, say $Ax^2 + Bx + C$. By the quadratic formula, the roots of this factor are complex conjugates (since they are not real), and so they are a and $\bar{a}$. Since this quadratic divides f, $\bar{a}$ is also a root of f.

*12. *If* $\mathbf{E}$ *is a finite extension of* $\mathbf{F}$, *show that there are only a finite number of intermediate fields* $\mathbf{K}$, $\mathbf{E} \supset \mathbf{K} \supset \mathbf{F}$. *(This may be surprising.)*

By Theorem 18, we may write $\mathbf{E} = \mathbf{F}(a)$ for some a algebraic over $\mathbf{F}$. Let f be a minimal polynomial for a over $\mathbf{F}$. Since a is certainly algebraic over any intermediate field $\mathbf{K}$, for each such field there exists a unique monic minimal polynomial g for a. By Problem 6 of Section 2.1, g divides f. Furthermore, if $\mathbf{K}_1$ and $\mathbf{K}_2$ are distinct intermediate fields and g_1 and g_2 the corresponding minimal polynomials for a, we will show that $g_1 \neq g_2$. For if $g_1 = g_2$, define $\mathbf{K}$ to be the multiple (algebraic) extension of $\mathbf{F}$ by the coefficients of g_1 and g_2. Since $\deg_{\mathbf{K}} a = \deg_{\mathbf{K}_1} a$ and $\mathbf{E} = \mathbf{K}(a) = \mathbf{K}_1(a)$, by Theorem 17, $[\mathbf{E} : \mathbf{K}] = [\mathbf{E} : \mathbf{K}_1]$. By Theorem 19 then, $[\mathbf{K} : \mathbf{F}] = [\mathbf{K}_1 : \mathbf{F}]$. Since in addition $\mathbf{K}_1 \supset \mathbf{K}$, we can conclude that $\mathbf{K} = \mathbf{K}_1$. Similarly, $\mathbf{K} = \mathbf{K}_2$, and so $\mathbf{K}_1 = \mathbf{K}_2$.

Consequently, distinct intermediate fields have distinct g's. But f obviously has only a finite number of monic factors in $\mathbf{E}[x]$, and so there can only be a finite number of intermediate fields.

13. *Suppose that* $\mathbf{E} = \mathbf{F}(r)$, *where* r *is a root of an irreducible polynomial* f *over* $\mathbf{F}$. *If* $\tilde{r}$ *is another root of* f *and if* $\tilde{r} \in \mathbf{E}$, *show that* $\mathbf{E} = \mathbf{F}(\tilde{r})$.

Since $\tilde{r} \in \mathbf{E}$, we can write $\mathbf{F} \subset \mathbf{F}(\tilde{r}) \subset \mathbf{E}$. Since $\deg_{\mathbf{F}} r = \deg_{\mathbf{F}} \tilde{r}$, Theorem 17 implies that $[\mathbf{E} : \mathbf{F}] = [\mathbf{F}(r) : \mathbf{F}] = [\mathbf{F}(\tilde{r}) : \mathbf{F}]$. By Theorem 18, $[\mathbf{E} : \mathbf{F}] = [\mathbf{E} : \mathbf{F}(\tilde{r})][\mathbf{F}(\tilde{r}) : \mathbf{F}]$, and so $[\mathbf{E} : \mathbf{F}(\tilde{r})] = 1$. By Problem 3, $\mathbf{E} = \mathbf{F}(\tilde{r})$.

Section 2.3

1. *Use the techniques of this section to show that it is not possible to divide an arbitrary angle into five equal parts by the usual rules of construction.*

The relevant equation was derived in Problem 3 of Section 1.3. With $x = 2\cos\theta$, we have $x^5 - 5x^3 + 5x - 2\cos 5\theta = 0$. If we take 5θ such that $\cos 5\theta = 5/6$, which angle is constructible since $5/6 \in \mathbf{Q}$, we see that the resulting polynomial is irreducible. (Multiply through by 3 and apply the Eisenstein criterion with $p = 5$.) Thus $2 \cos \theta$, being a root of this, is not constructible. But if this particular constructible 5θ were able to be divided into five equal parts, $\cos \theta$ and hence $2 \cos \theta$ would have to be constructible.

*2. *Let* $S = \{\theta \mid \theta$ *is a trisectible angle*$\}$. *Prove that* S *is countable*. (*Keep in mind the distinction drawn at the end of Section* 1.3.)

Since the set of algebraic numbers is countable, it follows that the set of angles $3\theta'$ for which $2 \cos 3\theta'$ is transcendental has for its complement a countable set. We show that for each such $3\theta'$, $3\theta'$ is not trisectible. If it were, by an obvious extension of Theorem 1, there would have to be a sequence of fields $\mathbf{Q}(2 \cos 3\theta') = \mathbf{F}_0 \subset \mathbf{F}_1 \subset \cdots \subset \mathbf{F}_N$ such that $2 \cos \theta' \in \mathbf{F}_N$ and such that each $\mathbf{F}_{j+1}$ is a quadratic extension of $\mathbf{F}_j$. Letting $a = 2 \cos 3\theta'$, at least one root of $x^3 - 3x - a$, namely $2 \cos \theta'$, would have to have some power of 2 for its degree over $\mathbf{Q}(a)$. For this to happen, $x^3 - 3x - a$ would have to be reducible over $\mathbf{Q}(a)$; otherwise each of its roots would have degree 3 over $\mathbf{Q}(a)$. But if $x^3 - 3x - a$ is reducible over $\mathbf{Q}(a)$, at least one of its factors must be linear, and so it would have a root in $\mathbf{Q}(a)$. Since every element in $\mathbf{Q}(a)$ has the form $f(a)/g(a)$ where f and g are polynomials over $\mathbf{Q}$ evaluated at a, we would thus have for some f and g

$$[f(a)/g(a)]^3 - 3[f(a)/g(a)] - a = 0,$$

and hence

$$[f(a)]^3 - 3[f(a)][g(a)]^2 - a[g(a)]^3 = 0.$$

This would give a as a root of the polynomial

$f^3 - 3fg^2 - xg^3 \in \mathbf{Q}[x]$, a contradiction as long as we can show that this polynomial is not the 0 polynomial. Let $n = \deg f$ and $m = \deg g$. (Obviously, neither can be the 0 polynomial.) The three terms have respective degrees $3n$, $n + 2m$, and $3m + 1$. If $n > m$, there is a single term of highest degree $3n$. If $n \leqslant m$, there is a single term of highest degree $3m + 1$. In either case, the terms cannot all cancel and so the polynomial is nontrivial.

Section 2.4

1. *Determine the cube roots of unity by a geometrical analysis.*

Dividing the unit circle into three equal parts, beginning at (1, 0), gives for the other two points (cos 120°, sin 120°) and (cos 240°, sin 240°). Evaluating these trigonometric functions, we get the points $(-1/2, +\sqrt{3}/2)$ and $(-1/2, -\sqrt{3}/2)$, which correspond to the complex numbers $-1/2 + i\sqrt{3}/2$ and $-1/2 - i\sqrt{3}/2$.

2. *Evaluate $\phi(k)$ for $k = 1, 2, \ldots, 15$.*

$\phi(1) = 1$, $\phi(2) = 1$, $\phi(3) = 2$, $\phi(4) = 2$, $\phi(5) = 4$, $\phi(6) = 2$, $\phi(7) = 6$, $\phi(8) = 4$, $\phi(9) = 6$, $\phi(10) = 4$, $\phi(11) = 10$, $\phi(12) = 4$, $\phi(13) = 12$, $\phi(14) = 6$, $\phi(15) = 8$.

3. *Show that $\phi(p) = p - 1$ for p a prime.*

Since p is prime, every number from 1 through $p - 1$ is relatively prime to p.

4. *Determine $\phi(p^2)$ for p a prime and n a positive integer.*

Of the p^n numbers from 1 to p^n, only $p, 2p, \ldots, p^{n-1}p$ are not relatively prime to p. Since there are p^{n-1} of these numbers, $\phi(p^n) = p^n - p^{n-1} = p^{n-1}(p - 1)$.

5. *Would it make sense to discuss "primitive" nth roots of 2?*

No. If $z^m = 2$, $1 \leqslant m \leqslant n$, then $|z|^m = 2$, which for a fixed z can only happen for $m = n$. Thus although every root might be termed "primitive", because of this fact the term would lose any usefulness. Moreover, the basic properties of primitive roots of unity would not carry over.

6. *How many primitive* 20*th roots of unity are there?*

Eight, since $\phi(20) = 8$.

7. *Describe the location of the seventh roots of* $1 - i$ *in the complex plane.*

An argument of $1 - i$ is $-\pi/4$. Take one seventh of this to get an argument of one root. Since $|1 - i| = \sqrt{2}$, the modulus of any root is $\sqrt[14]{2}$. Thus the roots are equally spaced on the circle $x^2 + y^2 = \sqrt[7]{2}$, beginning with the point corresponding to $\theta = -\pi/28$.

8. *Show that if a is an nth root of unity, then a is a primitive mth root of unity for some m such that $m \mid n$.*

It is obvious that a is a primitive mth root of unity for some m such that $1 \leqslant m \leqslant n$. Just let m be the *smallest* positive integral exponent such that $a^m = 1$. To show that $m \mid n$, use the division algorithm for $\mathbf{Z}$ to write $n = mq + r$, where $0 \leqslant r < m$. We simply want to show that $r = 0$, which we do by showing that $a^r = 1$. To do this, we calculate $a^r = a^{n-mq} = a^n/a^{mq} = 1/(a^m)^q = 1/1 = 1$.

9. *Show that $\sum_{d \mid n} \phi(d) = n$, where the sum is over all positive integers d which divide n. Verify this result by direct calculation for the case $n = 12$.*

Both sides represent the number of nth roots of unity. For the right side this is obvious. For the left, this follows from the previous problem, noting that the m there is unique. For $n = 12$, $\phi(1) + \phi(2) + \phi(3) + \phi(4) + \phi(6) + \phi(12) = 1 + 1 + 2 + 2 + 2 + 4 = 12$.

10. *Determine the degree over* **Q** *of the primitive nth roots of unity for* $n = p$ *and* $n = p^2$, *where p is a prime.*

For $n = p$, they are the roots of $(x^p - 1)/(x - 1) = x^{p-1} + x^{p-2} + \cdots + 1$, which was shown in Problem 8 of Section 2.1 to be irreducible over **Q**. Hence in this case they have degree $p - 1$.

For $n = p^2$, they are the roots of $(x^{p^2} - 1)/(x^p - 1) = x^{p(p-1)} + x^{p(p-2)} + \cdots + 1$. This equation is irreducible, as can be seen by means of the change of variable $x = u + 1$. For,

$$\sum_{k=1}^{p} (u+1)^{p(p-k)} = \sum_{k=1}^{p} \left[(u+1)^p\right]^{p-k}$$

$$= \sum_{k=1}^{p} \left[u^p + 1 + pg(u)\right]^{p-k}$$

$$= \sum_{k=1}^{p} \left[(u^p + 1)^{p-k} + ph_k(u)\right]$$

where each $h_k(u)$ is a polynomial of degree $(p-1)(p-k)$ with integral coefficients. Using the formula for the sum of a finite geometric progression,

$$\sum_{k=1}^{p} (u+1)^{p(p-k)} = \sum_{k=1}^{p} (u^p + 1)^{p-k} + pH(u)$$

$$= \frac{(u^p + 1)^p - 1}{u^p} + pH(u)$$

$$= u^{p(p-1)} + pG(u)$$

for some polynomial G of degree $(p-1)^2$. To apply the Eisenstein criterion, it simply remains to show that the constant term of the entire expression is not divisible by p^2. Although we could trace this value through the computations, it is easier just to plug $u = 0$ into the original sum, which gives the constant term $= p$. Now by the Eisenstein criterion the polynomial in u, and hence in x, is irreducible. Therefore the degree over $\mathbf{Q}$ of the primitive p^2 roots of unity is $p(p-1)$.

11. *If k and n are relatively prime positive integers, show that $\sum_i r_i^k = 0$, where the r_i's are the n distinct nth roots of unity.*

The numbers r_i^k are obviously roots of unity themselves, since $(r_i^k)^n = (r_i^n)^k = 1^k = 1$. They are also distinct, for if $r_i^k = r_j^k$, then $(r_i/r_j)^k = 1$. This combined with $(r_i/r_j)^n = 1$ would make r_i/r_j a primitive mth root of unity for some m dividing both k and n. Hence $m = 1$ and $r_i = r_j$. Therefore $\sum r_i^k = \sum r_i$, which equals the negative of the coefficient of x^{n-1} in the polynomial $x^n - 1$. (Section 1.5.) This coefficient is 0.

*12. *Determine the degree over $\mathbf{Q}$ of each of the primitive 20th roots of unity. (This is not the same question as in Problem 6.)*

The degree is 8. We begin by factoring $x^{20} - 1 = (x^{10} - 1)(x^{10} + 1)$; the primitive 20th roots must be roots of the second factor. Further, $x^{10} + 1 = (x^2 + 1)(x^8 - x^6 + x^4 - x^2 + 1)$, from which it is clear that the primitive 20th roots, of which there are precisely 8 (Problem 6), are the roots of $x^8 - x^6 + x^4 - x^2 + 1$. However, we don't yet know whether this can be factored further over $\mathbf{Q}$, and the Eisenstein criterion does not apply even with the kinds of variable changes used in Section 2.1 and in Problem 10.

Since any primitive 20th root of unity may be written as a power of any other, they all have the same degree over $\mathbf{Q}$ and in fact $\mathbf{Q}(\omega_1) = \mathbf{Q}(\omega_2)$. Consequently, if $x^8 - x^6 + x^4 - x^2 + 1$ can be factored in $\mathbf{Q}[x]$, its irreducible factors must all have the same degree. Hence there are either 8 linear factors, 4 irreducible quadratic factors, or 2 irreducible quartic factors. In any of these cases, this polynomial can certainly be written as the product of 2 quartic factors (although not necessarily irreducible ones). That is, we can write

$$x^8 - x^6 + x^4 - x^2 + 1 = \left[ax^4 + bx^3 + cx^2 + dx + e\right] \times \left[Ax^4 + Bx^3 + Cx^2 + Dx + E\right]$$

for some rational and, by the Lemma of Gauss, integral coefficients $a, b, c, d, e, A, B, C, D, E$. By comparing corresponding coefficients we shall be able to obtain contradiction, thus showing that the primitive 20th roots of unity have degree 8 over $\mathbf{Q}$. In particular, taking $a = A = 1$ without loss of generality, we can conclude rather easily that $b = -B, d = -D, c = C$, and $e = E = \pm 1$. Further, from $-1 = 2C - B^2$ it follows that B is odd and that, since $2C = (B-1)(B+1)$, C is even. At the same time, however, from $1 = 2E - 2BD + C^2$ it follows that C is odd. This gives the desired contradiction.

(It turns out that the $\phi(n)$ primitive nth roots of unity always satisfy a single irreducible polynomial over $\mathbf{Q}$ but it is rather difficult to prove this. See Problem 5 of Section 2.7.)

*13. *Show that the degree over* $\mathbf{Q}$ *of each of the primitive nth roots of unity when* $n = p^m$, *a power of a prime, is exactly* $\phi(n)$.

The roots in question are precisely the roots of the polynomial

$$f(x)=\frac{x^n-1}{x^{p^{m-1}}-1}=x^{(p-1)p^{m-1}}+x^{(p-2)p^{m-1}}$$
$$+\cdots+x^{p^{m-1}}+1$$

whose degree is $\phi(n)$. (Cf. Problem 4.) Thus it suffices to show that f is irreducible over $\mathbf{Q}$. Note that $f(1)=p$. If there were a nontrivial factorization $f(x)=g(x)h(x)$, by the Lemma of Gauss we could assume that g and h had integral coefficients. In this case, when $x=1$, one of them has value ± 1 and the other $\pm p$. In fact, without loss of generality we may assume that $g(1)=1$. Now g has at least one root, which must be a primitive nth root of unity. Call it ω. The powers of ω from ω^1 to ω^{n-1} must include all the primitive nth roots of unity, along with other nth roots of unity. Therefore the polynomial

$$G(x)=g(x)g(x^2)\cdots g(x^{n-1})$$

contains among its roots all the roots of f, and thus is divisible by f. So we can write

$$\frac{G(x)}{f(x)}=q(x)$$

for some polynomial q. Since f is primitive, Problem 11 of Section 2.1 implies that q has integral coefficients, so that $q(1)$ must be an integer. However, we find that

$$q(1)=\frac{G(1)}{f(1)}=\frac{1}{p},$$

a contradiction which shows that f could not have been reducible. (The result of this problem holds for all n, not just powers of a prime. See Problem 5, Section 2.7.)

Section 2.5

1. *Give a proof of the multiplication and division parts of Lemma 22a using the polar representation of a_1 and a_2.*

If $a_1 = |a_1|e^{i\theta_1}$ and $a_2 = |a_2|e^{i\theta_2}$, then $a_1a_2 = |a_1|\,|a_2|e^{i(\theta_1+\theta_2)}$, whose modulus is constructible since it is the product of constructible numbers and whose argument is constructible since it is obviously possible to add two angles. Since $a_1/a_2 = (|a_1|/|a_2|)e^{i(\theta_1-\theta_2)}$, the argument here is very similar.

2. *Give a proof of Lemma 22b which does not use the polar representation of a.*

Letting $a = b + ic$ and $k = r + is$, we simply solve for r and s in terms of b and c. We have $k^2 = (r + is)(r + is) = (r^2 - s^2) + i(2rs) = b + ic$, so that $r^2 - s^2 = b$ and $2rs = c$. Hence $r = c/2s$, so that $c^2 - 4s^4 = 4bs^2$, from which we can solve for s by means of two quadratic equations, the steps therein corresponding to admissible constructions. Since $r = \pm\sqrt{s^2 + b}$, r is also constructible.

3. *Give an analytic proof of Lemma 24a.*

If ω is a primitive nth root of unity, then for $n = mk$, ω^k is easily seen to be a primitive mth root. Since by the hypothesis ω is constructible, so too is ω^k.

4. *Prove that the regular pentagon is constructible.*

The primitive fifth roots of unity satisfy the irreducible polynomial equation $(x^5 - 1)/(x - 1) = x^4 + x^3 + x^2 + x + 1 = 0$. This is a "reciprocal equation", meaning that the reciprocal of any root is also a root; for, dividing by x^4, we obtain $1 + (1/x) + (1/x)^2 + (1/x)^3 + (1/x)^4 = 0$. In such cases the change of variables $u = x + (1/x)$

halves the degree (which must be even to start or else the equation would be reducible, having $+1$ or -1 for a solution). Dividing the original by x^2, we have

$$x^2 + x + 1 + x^{-1} + x^{-2} = 0.$$

The substitutions,

$$u = x + x^{-1},$$
$$u^2 = x^2 + 2 + x^{-2},$$

lead to $u^2 + u - 1 = 0$. Thus u can be obtained by the solution of a quadratic, and from u, x can be obtained likewise. Hence the roots are constructible. (By carrying through the computations, one could actually describe the construction.)

5. *Is it possible to divide an angle of* $60°$ *into five equal parts*?

Yes. From the previous problem $72° = 360°/5$ is constructible, and $12° = 72° - 60°$.

Section 2.6

1. *Use congruences to reformulate and prove the theorem that every perfect square is of the form* $4n$ *or* $4n + 1$.

For any a, $a \equiv 0$, 1, 2, or 3 (mod 4). In these cases $a^2 \equiv 0, 1, 0, 1 \pmod 4$ respectively, which is equivalent to the theorem.

2. *If* $a \not\equiv 0 \pmod n$ *and* $b \not\equiv 0 \pmod n$, *does it follow that* $ab \not\equiv 0 \pmod n$?

No, $3 \not\equiv 0 \pmod 6$ and $2 \not\equiv 0 \pmod 6$, but $6 \equiv 0$ (mod 6). This is equivalent to the fact that n may divide a product of two numbers without dividing either one. Of course, if n is prime, this can no longer happen.

3. *If a and n are relatively prime, show that there exists an integer b such that $ab \equiv 1 \pmod{n}$.*

The number $b = a^{\phi(n)-1}$ will do, by Fermat's Theorem. For a direct proof, note that in the list $a, a^2, \ldots, a^{n+1}$, some two entries must be congruent, for each is congruent to one of the n numbers $0, 1, \ldots, n-1$. But if $a^j \equiv a^k \pmod{n}$, with, say, $k > j$, then $1 \equiv a^{k-j} \equiv a \cdot a^{k-j-1} \pmod{n}$, so that we can take $b = a^{k-j-1}$.

4. *With the obvious definition, does there always exist a primitive root modulo n, when n is not prime?*

No, and $n = 8$ is the smallest counterexample. In general, we would want a to have order $\phi(n)$ modulo n. Here we have $\phi(8) = 4$, but for every a, $1 \leqslant a \leqslant 7$, with $(a, 8) = 1$, there is a lower power congruent to 1. The a's are: 1, 3, 5, and 7; and by computation we find that $1^1 \equiv 3^2 \equiv 5^2 \equiv 7^2 \equiv 1 \pmod{8}$.

In number theory it is shown that the only numbers n having primitive roots modulo n are those of the form p^k, $2p^k$, 2, and 4, where p is an odd prime.

5. *Find all the primitive roots modulo* 7.

We are looking for numbers of order $\phi(7) = 6$ modulo 7. By Theorem 27 there are $\phi(p-1) = \phi(6) = 2$ of them. By trial and error we find two such to be 3 and 5. As alternatives, of course, we could take any numbers congruent to these modulo 7.

6. *How many primitive roots modulo* 17 *are there? Find one.*

There are $\phi(16) = 8$ of them. They are: 3, 5, 6, 7, 10, 11, 12, and 14.

Section 2.7

1. *Carry through the proof of Theorem 28 for the case $p = 5$, and deduce from it an algorithm for the construction of the regular pentagon.*

Let ω be a primitive fifth root of unity. A primitive root modulo 5 is $g = 2$, and so we can list the four primitive fifth roots of unity as: $\omega^2, \omega^4, \omega^3, \omega^1$. Taking in particular $\omega = e^{2\pi i/5}$, these are sketched in Figure 17. Proceeding as in the proof, $\eta_1 = \omega^2 + \omega^3$ and $\eta_2 = \omega^4 + \omega^1$, so that $\eta_1 + \eta_2 = -1$ and $\eta_1\eta_2 = -1$. Thus η_1 and η_2 are the roots of $x^2 + x - 1$. Since, as is clear from the diagram, $\eta_2 > \eta_1$, we have $\eta_1 = (-1 - \sqrt{5})/2$ and $\eta_2 = (-1 + \sqrt{5})/2$. Two of the 2-periods are just ω^4 and ω^1. Their sum is η_2 and their product is 1. Thus we see that they are the roots of

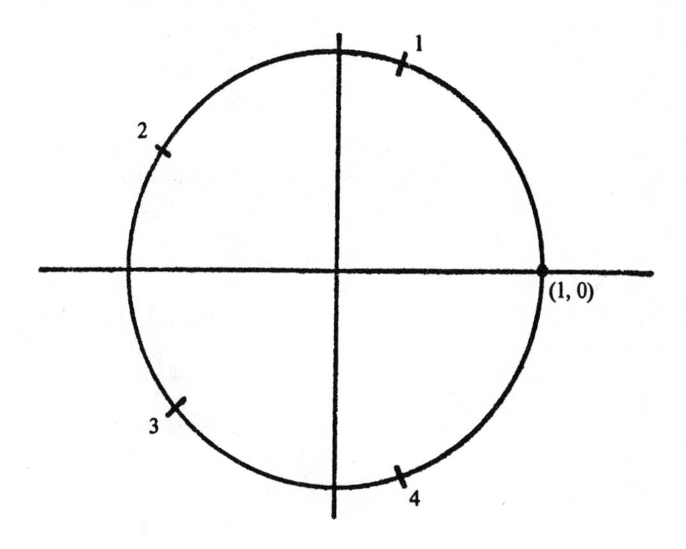

Figure 17

$x^2 - \eta_2 x + 1$, from which it follows that

$$\omega = \frac{\eta_2 + \sqrt{\eta_2^2 - 4}}{2}.$$

Thus the algorithm is to construct successively η_2 and then ω.

2. *Generalize the method of this section to show that the primitive seventh roots of unity may be obtained by the successive solution of polynomials of degrees* 2 *and* 3. (*You are not asked to solve these lower degree polynomials.*)

The number $g = 3$ is a primitive root modulo 7. Letting ω be a primitive seventh root of unity, we can list all the primitive roots in the usual way as $\omega^3, \omega^2, \omega^6, \omega^4, \omega^5, \omega^1$. We break this set into three groups, taking every third one, and thus obtain periods

$$\begin{aligned}\eta_1 &= \omega^3 + \omega^4,\\ \eta_2 &= \omega^2 + \omega^5,\\ \eta_3 &= \omega^6 + \omega^1.\end{aligned}$$

We shall find a cubic with these three periods as its roots by evaluating the three elementary symmetric functions of three variables at these values. That is,

$$\eta_1 + \eta_2 + \eta_3 = -1$$

$$\begin{aligned}\eta_1\eta_2 + \eta_1\eta_3 + \eta_2\eta_3 &= [\omega^5 + \omega^1 + \omega^6 + \omega^2]\\ &\quad + [\omega^2 + \omega^4 + \omega^3 + \omega^5]\\ &\quad + [\omega^1 + \omega^3 + \omega^4 + \omega^6]\\ &= -2\end{aligned}$$

$$\begin{aligned}\eta_1\eta_2\eta_3 &= \omega^4 + \omega^0 + \omega^5 + \omega^1 + \omega^6 + \omega^2 + \omega^0 + \omega^3\\ &= 2 + (-1) = 1.\end{aligned}$$

Consequently, η_1, η_2, and η_3 are the roots of $x^3 + x^2 - 2x - 1$. Now each of these 1-periods may be decomposed into two 2-periods, each of which is a power of ω. These can be found as the solutions of quadratics. For example, $\omega^3 + \omega^4 = \eta_1$ and $(\omega^3)(\omega^4) = 1$, so that ω^3 and ω^4 are the roots of $x^2 - \eta_1 x + 1$. Similarly for the others.

(Actually, for any prime p, this method reduces the solution of the pth cyclotomic polynomial to the solution of equations whose degrees are the prime factors of $p - 1$. It can also be shown that all the resulting equations can actually be solved explicitly, but this topic is best addressed in the more general context of Chapter 3.)

3. *If a is a primitive mth root of unity and if b is a primitive nth root of unity, and if $(m, n) = 1$, show that ab is a primitive mnth root of unity. Use this fact to give an alternate proof of Lemma 28a.*

Since $a^m = 1$ and $b^n = 1$, it is clear that $(ab)^{mn} = 1$; we simply need to show that mn is the minimal such positive exponent. Suppose that for some $k \geqslant 1$, $(ab)^k = 1$; we will show that $(mn) \mid k$. Since $a^k = b^{-k}$, we have $1 = a^{km} = b^{-km}$, and hence, by the same reasoning as in the proof of Theorem 21, $n \mid (-km)$. Since $(m, n) = 1$, in fact $n \mid k$. Similarly, we get $m \mid k$. But since $(m, n) = 1$, it follows that $(mn) \mid k$. To complete an alternate proof of Lemma 28a, we note that there we are essentially given the constructibility of a and b. By Lemma 22a, ab is constructible, and hence so is the regular mn-gon.

*4. *Find a necessary and sufficient condition on a rational number α such that an angle of α degrees is constructible.*

If α is written M/N in lowest terms, then for α to be constructible it is necessary and sufficient that 3 divide M

and that N have the form $2^k p_1 p_2 \cdots p_m$ where the p_i's are distinct Fermat primes other than 3 or 5. (Of course we may have $m = 0$, in which case N is just a power of 2.) The proof consists essentially of an application of Theorems 24 and 28 (on the constructibility of regular n-gons). Important tools are the Euclidean algorithm and the observation that an angle of 3° is constructible (the central angle of a pentagon is 72°; subtract 60° and bisect twice) but that 1° is not (for then the multiple 20° would be, which was shown impossible in Section 1.3). The degree symbol will be omitted in what follows.

(*Necessity.*) If M/N is constructible, then so is M. If $(M, 3) = 1$, then we could write $Mr_1 + 3s_1 = 1$ for integers r_1 and s_1 and so 1 would be constructible, which is impossible. Hence $3 \mid M$. For convenience we will write $M = 3Q$; of course $(Q, N) = 1$. But then $Qr_2 + Ns_2 = 1$ for integers r_2 and s_2, and so $3Qr_2/N + 3s_2 = 3/N$. Since the first term on the left is a multiple of α and the second is a multiple of 3, we see that $3/N$ is constructible. Therefore the multiple $360/N$ is constructible, from which Theorem 24 implies that N has the form $2^k p_1 p_2 \cdots p_m$. It remains to show that none of the primes can be 3 or 5. The case of 3 is excluded since $3 \mid M$ and $(M, N) = 1$. For the case $5 \mid N$, it would follow that $3/5$ is constructible since it is a multiple of $3/\text{N}$. But then $24(3/5) = 72/5$ would be constructible, which is impossible since this is the central angle of a regular 25-gon, which is not constructible.

(*Sufficiency.*) Let α have the given form. We know that $360/(2^k \cdot 3 \cdot 5 \cdot p_1 \cdots p_m)$ is constructible by Theorem 28. This is just $(3 \cdot 8)/(2^k p_1 p_2 \cdots p_m)$ which can be bisected thrice and then multiplied by Q to give the angle α.

*5. *Show that all the $\phi(n)$ primitive nth roots of unity are the roots of a single irreducible polynomial over* $\mathbf{Q}$ *which is monic and has integral coefficients. Use this result to give an alternate proof of Theorem 24. (Hint: This problem does not depend specifically on the results of this section. It was not included in an earlier section simply so the reader would have more experience with some of the material before attempting it. Special cases were encountered in the problems in Section* 2.4.)

The polynomial $x^n - 1$ can be factored into irreducible polynomials; by Gauss' Lemma they may be taken to have integral coefficients, in which case they must also be monic. If ω is a primitive nth root of unity, then the factor having ω as a root will be called f. We want to show that this irreducible f has exactly all the $\phi(n)$ primitive nth roots of unity as its roots, and thus that it is the nth cyclotomic polynomial.

Consider the polynomial $f_k(x) = f(x^k)$, which is also monic and has integral coefficients. By the division algorithm we can write $f_k = q_k f + r_k$ where either $r_k = 0$ or $\deg r_k < \deg f$. In addition, since f is monic, Problem 12 of Section 2.1 implies that q_k and r_k have integral coefficients. If we do the division with some other f_j, $j \equiv k \pmod{n}$, the resulting r_j is identical with r_k. To see this note that $(f_k - f_j)(\omega) = f(\omega^k) - f(\omega^j) = 0$ and so $f \mid (f_k - f_j)$. Thus the remainder when we divide $f_k - f_j$ by f is 0; but this remainder is simply $r_k - r_j$. In conclusion, as k ranges through all the positive integers, there are only n different remainders $r_1, r_2, \ldots, r_n$ that we may obtain as above. Let M be an integer greater than the absolute value of every coefficient in every one of these r_i's. We claim that for every prime $p > M, f \mid f_p$. To see this, look at the polynomial f^p, f raised to the pth power. It is easy to see that it has the form $f^p = f_p + pg$ for some polynomial g with integral coefficients, and we can further write $g = sf + t$ for polynomials s and t with integral coefficients, $t = 0$ or $\deg t < \deg f$. But now we have two representations of $f_p : f_p = q_p f + r_p$, and $f_p = f^p - pg = f^p - psf - pt = (f^{p-1} - ps)f - pt$. By the uniqueness in the division algorithm it follows that $-pt = r_p$. But r_p equals one of r_1 through r_n, and so if $p > M$, the only way all its coefficients can be divisible by p is if $r_p = 0$. That is, for all $p > M, f \mid f_p$.

Every one of the primitive nth roots of unity has the form ω^k for some k, $1 \leqslant k \leqslant n-1$, such that $(k, n) = 1$. Of course if $j \equiv k \pmod{n}$, then $\omega^j = \omega^k$. Thus let us fix k as above and seek such a j for which we can prove that ω^j is a root of f. Let P denote the product of all primes less than or equal to M except those which divide k. We claim that $j = k + nP$ is an appropriate j. (Obviously $j \equiv k \pmod{n}$.) Every prime factor of j must be greater than M, for any smaller prime will be a factor of exactly one of k and nP. Thus the prime factorization $j = p_1 p_2 \cdots p_m$, possibly with repetitions, includes only primes greater than M. By our earlier result then, each p_i is such that $f \mid f_{p_i}$. Now we repeatedly apply this fact. Since ω is a root of f, it is a root of f_{p_1}, that is, ω^{p_1} is a root of f. But then ω^{p_1} is also a root of f_{p_2}, which implies that $(\omega^{p_1})^{p_2} = \omega^{p_1 p_2}$ is a root of f. Continuing in this fashion, we obtain that ω^j is a root of f, which completes the proof.

To prove Theorem 24 we want to determine for what values of n is $\phi(n)$ a power of 2. It follows easily from the definition of ϕ that if j and k are relatively prime, then $\phi(jk) = \phi(j)\phi(k)$. Therefore, if n has the prime power decomposition $n = p_1^{m_1} p_2^{m_2} \cdots p_N^{m_N}$, then $\phi(n)$

$=\prod_i p_i^{m_i-1}(p_i-1)$. Thus if p_i is an odd prime, we must have both $m_i=1$ and p_i-1 equal to a power of 2. This is exactly what we want.

Section 3.1

1. *Show that the general cubic equation* $ax^3+bx^2+cx+d=0$ *can be solved by radicals. (Hint: Without loss of generality, take* $a=1$. *Make the successive substitutions* $x=y+L$ *and* $y=z+K/z$ *for appropriate choices of* L *and* K.)

Without loss of generality we can take $a=1$, for otherwise we could divide through by it. (Since it is a cubic, $a\neq 0$. If it were 0, we would have at most a quadratic, which is solvable.) The substitution $x=y+L$ leads to the equation

$$y^3+y^2[3L+b]+y[3L^2+2bL+c]$$
$$+[L^3+bL^2+cL+d]=0.$$

The y^2 term drops out by taking $L=-b/3$. So the equation is now in the form

$$y^3+By+C=0.$$

If $C=0$, its solution is apparent. If $C\neq 0$, then $y=0$ is not a root of this equation. The substitution $y=z+K/z$ leads to

$$z^6+z^4[3K+B]+Cz^3+z^2[3K+B]K+K^3=0.$$

Two terms drop out by the choice $K=-B/3$, leaving

$$z^6+Cz^3+K^3=0,$$

or

$$(z^3)^2 + C(z^3) + K^3 = 0.$$

This is a quadratic in which we can solve for z^3. Thus z can also be obtained by radicals. From z it is easy to get back to x.

(There remains a question of which square root and which cube root to take in the above in order to get a (non-extraneous) solution to the equation. This question goes beyond that of whether the equation is solvable by radicals, but in any case for a given equation the actual roots can be determined by trying all six combinations.)

2. *What is the degree of the splitting field of* $x^3 - 2$ *as an extension of* $\mathbf{Q}$? *Find a basis for this extension.*

If ω is a primitive cube root of unity, then the splitting field is given by $\mathbf{E} = \mathbf{Q}(\sqrt[3]{2}, \omega\sqrt[3]{2}, \omega^2\sqrt[3]{2})$. We can build up to $\mathbf{E}$ by two simple extensions of $\mathbf{Q}$. Let $\mathbf{F}_1 = \mathbf{Q}(\omega)$, which is an extension of degree 2 since ω is a root of $(x^3 - 1)/(x - 1) = x^2 + x + 1$. A basis for this extension is $\{1, \omega\}$. Since the irreducible (over $\mathbf{Q}$) polynomial $x^3 - 2$ cannot have a root in the quadratic extension $\mathbf{F}_1$, it is irreducible over $\mathbf{F}_1$, so that with $\mathbf{F}_2 = \mathbf{F}_1(\sqrt[3]{2})$, a basis for $\mathbf{F}_2$ over $\mathbf{F}_1$ is $\{1, \sqrt[3]{2}, \sqrt[3]{4}\}$. Clearly, $\mathbf{F}_2 = \mathbf{E}$, and as in the proof of Theorem 19, we may obtain a basis for $\mathbf{E}$ over $\mathbf{Q}$ by multiplying the elements of the bases of the two successive extensions to obtain $\{1, \sqrt[3]{2}, \sqrt[3]{4}, \omega, \omega\sqrt[3]{2}, \omega\sqrt[3]{4}\}$. Thus the degree of $\mathbf{E}$ over $\mathbf{Q}$ is 6.

3. *Find an irreducible polynomial over* $\mathbf{Q}$ *whose splitting field has degree less than* $n!$ *over* $\mathbf{Q}$, *where* n *is the degree of the polynomial.* (*Cf. Problem* 8 *of Section* 2.2.)

One such polynomial is $x^4 - 2$. If ω is a primitive 4th root of unity, all its roots are given by $\{\sqrt[4]{2}, \omega^1\sqrt[4]{2},$

$\omega^2 \sqrt[4]{2}$, $\omega^3 \sqrt[4]{2}$ }. The splitting field **E** can be built up in two steps. Let $\mathbf{F}_1 = \mathbf{Q}(i)$, which is of degree 2 but contains ω (since $\omega = \pm i$). Let $\mathbf{F}_2 = \mathbf{E}$. Then $[\mathbf{E} : \mathbf{Q}] = [\mathbf{E} : \mathbf{F}_1][\mathbf{F}_1 : \mathbf{Q}] \leqslant 4 \cdot 2 = 8 < 4! = 24$. We note that $[\mathbf{E} : \mathbf{F}_1] \leqslant 4$ since any extension of $\mathbf{F}_1$ with a single root of $x^4 - 2$ contains all its roots.

As another example, consider the splitting field of $x^{p-1} + x^{p-2} + \cdots + x + 1$, which is irreducible and whose roots are the primitive pth roots of unity, where p is a prime. An extension by any root contains all the roots, and so the degree of the splitting field is just $p - 1$, precisely the degree of the polynomial.

*4. *Show that the general quartic equation* $ax^4 + bx^3 + cx^2 + dx + e = 0$ *may be solved by radicals.* (*Hint*: *Without loss of generality, take* $a = 1$. *Let* $y = x + (b/4)$, *and then try to write the resulting equation as a difference of squares, making use of some extra parameter.*)

Following the suggestions, we obtain for y an equation of the form $y^4 + Cy^2 + Dy + E = 0$. Adding and subtracting $y^2t + t^2/4$ on the left, where t is a parameter, we obtain $(y^2 + t/2)^2 - [y^2(t - C) - Dy + t^2/4 - E] = 0$. We can make the expression in the brackets a perfect square by choosing t so that its roots are equal, or equivalently, by making its discriminant $(-D)^2 - 4(t - C)(t^2/4 - E)$ equal to 0. This latter gives a cubic equation for t:

$$-4t^3 + Ct^2 + 4Et + D^2 - 4CE = 0.$$

By Problem 1, this can be solved in radicals. Let t_0 be one of the roots of this cubic. Then

$$y^2(t_0 - C) - Dy + t_0^2/4 - E$$
$$= (t_0 - C)\left(y - \frac{D}{2(t_0 - C)}\right)^2.$$

Therefore we obtain by factoring our difference of squares

$$0 = \left[y^2 + t_0/2 - \sqrt{t_0 - C}\left(y - \frac{D}{2(t_0 - C)} \right) \right]$$

$$\times \left[y^2 + t_0/2 + \sqrt{t_0 - C}\left(y - \frac{D}{2(t_0 - C)} \right) \right].$$

Setting each factor equal to 0, we can obtain y in radicals. The question of which roots to take in different parts of this solution can be analyzed, but we do not do it here. The above calculation shows completely that any x that is a solution can be expressed by radicals. (The cubic we encountered in t is called the *resolvent cubic*.)

5. *Show that there exists a number which has degree* 4 *over* **Q** *but which is not constructible.* (*Hint*: *Consider the roots of* $f(x) = 4x^4 - 4x + 3$.)

We first divide the suggested polynomial by 4. From the calculations in the previous solution, we see that the resolvent cubic for f is $-t^3 + 3t + 1$, which is irreducible. Thus none of its roots can be constructed. However, in the notation of the previous problem, $\sqrt{t_0 + C} = \sqrt{t_0} = y_1 + y_2 = x_1 + x_2$, where x_1 and x_2 are roots of the original polynomial. Similarly, $\sqrt{t_0} = -x_3 - x_4$, where x_3 and x_4 are the other roots of f. Thus if all of the roots of f were constructible, t_0 would also be constructible. Since t_0 is not constructible, neither is some root of f. Since f has no rational roots, if it were reducible it would be as a product of quadratics over **Q**, and thus all the roots would be constructible. Since they are not, f is irreducible, and so each root has degree 4 over **Q**. (It can be shown that none of the roots of f are constructible.)

Section 3.2

1. *Suppose ϕ is a mapping from a field* **F** *to itself such that ϕ is one-to-one, onto, and preserves addition and multiplication (i.e., $\phi(a+b)=\phi(a)+\phi(b)$ and $\phi(ab) = \phi(a)\phi(b)$ for all $a, b \in \mathbf{F}$). Show that $\phi(0)=0$, $\phi(1)=1$, and that ϕ is an automorphism of* **F**.

Since $\phi(0)+\phi(0)=\phi(0+0)=\phi(0)$ we conclude that $\phi(0)=0$. Similarly, $\phi(1)\phi(1)=\phi(1\cdot 1)=\phi(1)$ so that $\phi(1)=1$, provided of course that $\phi(1)\neq 0$, which is true since ϕ is one-to-one and $\phi(0)=0$. To show that ϕ is an automorphism, it remains only to show that it preserves subtraction and division. We have $0=\phi(b-b)=\phi(b)+\phi(-b)$ so that $\phi(-b)=-\phi(b)$. Thus $\phi(a-b)=\phi(a)+\phi(-b)=\phi(a)-\phi(b)$. Similarly, for $b\neq 0$, $1=\phi(b/b) = \phi(b)\phi(1/b)$ so that $\phi(1/b)=1/\phi(b)$. Thus $\phi(a/b) = \phi(a)\phi(1/b)=\phi(a)/\phi(b)$.

2. *Let ϕ be an automorphism of a field* **F**. *Show that for every rational number q, $\phi(q)=q$. (That is, ϕ, restricted to* **Q**, *which is a subfield of* **F**, *is the identity mapping.)*

From the fact that $\phi(1)=1$ it follows from the additive property of ϕ that $\phi(n)=n$ for any positive integer n. For, $\phi(2)=\phi(1)+\phi(1)=2$ and the induction argument is obvious. For negative integers, the result follows from the fact that $\phi(-n)=-\phi(n)$, as shown in the previous problem. Thus, $\phi(m/n)=\phi(m)/\phi(n)=m/n$.

3. *Find all the automorphisms of* $\mathbf{Q}(\sqrt[3]{2})$.

There is only one, the identity mapping. For, if ϕ is an automorphism of $\mathbf{Q}(\sqrt[3]{2})$, $0=\phi(0)=\phi[(\sqrt[3]{2})^3-2] = [\phi(\sqrt[3]{2})]^3-\phi(2)=[\phi(\sqrt[3]{2})]^3-2$, by Problem 2. Thus $\phi(\sqrt[3]{2})$ is also a cube root of 2. But since $\mathbf{Q}(\sqrt[3]{2})\subset\mathbf{R}$, and since there is only one real cube root of 2, $\phi(\sqrt[3]{2}) = \sqrt[3]{2}$. If s is an arbitrary element of $\mathbf{Q}(\sqrt[3]{2})$, we know we can write $s = a + b(\sqrt[3]{2}) + c(\sqrt[3]{2})^2$, where a, b, and c are rational. But then $\phi(s) = s$, so $\phi = \text{id}$.

4. *Find all the automorphisms of* $\mathbf{Q}(\sqrt{2})$.

By the same reasoning as used in the last problem, $\phi(\sqrt{2})$ is a square root of 2. But in this case there are two possibilities, $\phi(\sqrt{2}) = \pm\sqrt{2}$. It is clear that each of these possibilities determines *at most* one automorphism, since $\mathbf{Q}(\sqrt{2}) = \{a + b\sqrt{2} \mid a, b \in \mathbf{Q}\}$. If $\phi(\sqrt{2}) = \sqrt{2}$, then $\phi(a + b\sqrt{2}) = a + b\sqrt{2}$. This yields the identity automorphism. If $\phi(\sqrt{2}) = -\sqrt{2}$, then $\phi(a + b\sqrt{2}) = a - b\sqrt{2}$. Is this an automorphism? We have to check the preservation of addition and multiplication. For addition, $\phi[(a + b\sqrt{2}) + (c + d\sqrt{2})] = \phi[(a + c) + (b + d)\sqrt{2}] = (a + c) - (b + d\sqrt{2}) = (a - b\sqrt{2}) + (c - d\sqrt{2}) = \phi(a + b\sqrt{2}) + \phi(c + d\sqrt{2})$. Similarly, it is easy to verify that ϕ preserves multiplication. Since ϕ is also one-to-one and onto, it is an automorphism. Thus there are two automorphisms of $\mathbf{Q}(\sqrt{2})$.

5. *Prove Theorem 29 using right cosets. (If H is a subgroup of G, a right coset of H is a set $\{\psi_i\phi\}$ where the ψ_i's range through H and ϕ is a fixed element of G.*

The proof may be repeated almost word for word, except that to show distinctness one needs to multiply through by inverses on the opposite side.

6. *Let G be any group, finite or not, and let H be a subgroup of G. If $\phi \in G$, then the set $\phi H = \{\phi\psi \mid \psi \in H\}$ is called the left coset of H by ϕ. Show that any two left cosets of G are either identical or disjoint.*

Let $\phi_1 H$ and $\phi_2 H$ be two left cosets. Suppose they have some element in common, call it τ. Then for some automorphisms $\psi_1, \psi_2 \in H$, we have $\phi_1\psi_1 = \tau = \phi_2\psi_2$, and thus $\phi_1 = \phi_2\psi_2\psi_1^{-1} \in \phi_2 H$. But this implies that $\phi_1 H \subset \phi_2 H$, for if $\phi_1\psi$ is an arbitrary element of $\phi_1 H$, we have $\phi_1\psi = \phi_2(\psi_2\psi_1^{-1}\psi) \in \phi_2 H$. Similarly, $\phi_2 H \subset \phi_1 H$. Thus if $\phi_1 H$ and $\phi_2 H$ have a single element in common, then $\phi_1 H = \phi_2 H$.

7. *If G is a group whose order is a prime p, how many subgroups does G have?*

It has two subgroups, G and $\{\text{id}\}$. By Lagrange's Theorem there can be no others.

8. *If G is a cyclic group of order n, determine exactly how many elements of G generate G.*

Let ψ be one generator of G. Then we claim that ψ^k is also a generator of G if and only if $(k, n) = 1$. For certainly, $(\psi^k)^n = \text{id}$, and the elements $\psi^k, (\psi^k)^2, \ldots, (\psi^k)^n$ will be distinct if and only if n is the least such positive power. Since ψ is a generator, $(\psi^k)^m = \text{id}$ if and only if $km \equiv 0 (\text{mod } n)$. If $(k, n) = 1$, then $n \mid m$, and so $m \geqslant n$. Conversely, if k and n have a common factor j, then $(\psi^k)^{n/j} = \text{id}$. The claim thus having been established, we simply observe that among $G = \{\psi, \psi^2, \ldots, \psi^n = \text{id}\}$ there are precisely $\phi(n)$ terms with exponents relatively prime to n, where here ϕ is the Euler ϕ-function.

9. *If G is a cyclic group of order n, and if $m \mid n$, show that G has a cyclic subgroup of order m.*

For any generator ϕ of G, we have $G = \{\phi, \phi^2, \ldots, \phi^n\}$. If $n/m = k$, define $\psi = \phi^k$. Then the cyclic subgroup generated by ψ clearly has order m, since

$\psi, \psi^2, \ldots, \psi^m$ are all distinct, being among the list $\phi, \phi^2, \ldots, \phi^n$; and also, $\psi^m = \phi^n = \text{id}$.

10. *Show that every subgroup of a cyclic group is cyclic.*

If we write $G = \{\phi, \phi^2, \phi^3, \ldots, \phi^n\}$, then H is of the form $\{\phi^i, \phi^j, \ldots, \phi^n = \text{id}\}$, where, without loss of generality, we may assume that $i < j < \cdots < n$. Then H is the cyclic subgroup generated by ϕ^i. To see this, observe that if $\phi^k \in H$, we can write $k = iq + r$, $0 \leqslant r < i$, and compute $\phi^r = \phi^k \phi^{-iq} \in H$. By the minimality of i, $r = 0$ and so k is a multiple of i. Therefore, $\phi^k = (\phi^i)^q$ and thus belongs to the cyclic subgroup generated by ϕ^i.

11. *If $|G| = p$, a prime, show that G is cyclic.*

Let $\phi \in G$, $\phi \neq \text{id}$. Then the cyclic subgroup generated by ϕ must have order p by Lagrange's Theorem.

12. *If H is a subgroup of prime index p in G, show that other than for H and G, there is no subgroup $\tilde{H}$ such that $H \subset \tilde{H} \subset G$.*

By Lagrange's Theorem, $|\tilde{H}| = k|H|$ for some integer $k \geqslant 1$. But $|G| = p|H|$ and $|\tilde{H}|$ divides $|G|$. Therefore $k = 1$ or p.

Section 3.3

1. *If p is a prime, show that the Galois group of $x^{p-1} + x^{p-2} + \cdots + 1 = (x^p - 1)/(x - 1)$ is cyclic and of order $p - 1$.*

The splitting field of the polynomial is $\mathbf{Q}(\omega)$, where ω is any primitive pth root of unity. By Theorem 31, the map

$\phi : \omega \to \omega^k$, for each k such that $1 \leqslant k \leqslant p-1$, determines a unique element of the Galois group, whose order is therefore $p-1$. (Recall from Problem 8 of Section 2.1 that this polynomial is irreducible over $\mathbf{Q}$.) If g is a primitive root modulo p, then the automorphism determined by the condition, $\phi : \omega \to \omega^g$, is a generator of the Galois group. For we have, $\phi^2 : \omega \to \omega^{g^2}$, $\phi^3 : \omega \to \omega^{g^3}$, etc., and the exponents will range through all congruence classes between 1 and $p-1$.

2. *Determine the order of* $G(\mathbf{E}/\mathbf{Q})$ *where* $\mathbf{E} = \mathbf{Q}(\sqrt{2}, \sqrt{3})$. (*Hint*: *See Problem* 7, *Section* 2.3, *and Problems* 1 *and* 2, *Section* 1.4).

By the problems cited, $\mathbf{E} = \mathbf{Q}(\sqrt{2} + \sqrt{3})$ (or something similar, depending on one's solutions), and $\sqrt{2} + \sqrt{3}$ has a minimal polynomial over $\mathbf{Q}$, $x^4 - 10x^2 + 1$. However, $\mathbf{E}$ is actually the splitting field of this polynomial, whose complete set of roots is $\sqrt{2} \pm \sqrt{3}$, $-\sqrt{2} \pm \sqrt{3}$. By Theorem 31, $|G(\mathbf{E}/\mathbf{Q})| = 4$.

3. *What is the order of the Galois group of* $x^5 - 1$?

It has the same splitting field, and hence the same group as, $x^4 + x^3 + x^2 + x + 1$, which by Problem 1 has a Galois group of order 4.

4. *Is there some polynomial* $f \in \mathbf{Q}[x]$ *such that* $\mathbf{Q}(\sqrt[3]{2})$ *is its splitting field*?

No. If $\mathbf{Q}(\sqrt[3]{2})$ were the splitting field of f, then f would have to have an irreducible factor in $\mathbf{Q}[x]$ of degree greater than 1. Thus not all the roots of f could be identical, since irreducible polynomials have distinct roots. But since they would all be in $\mathbf{Q}(\sqrt[3]{2})$, by Theorem

31 there would be more than one element of $G(\mathbf{Q}(\sqrt[3]{2})/\mathbf{Q})$, which, as shown in the text of this section, is impossible.

5. *If* $\phi \in G(\mathbf{E}/\mathbf{F})$, *where* **E** *is a finite extension of* **F**, *and if* **K** *is an intermediate field, show that the set* $\phi(\mathbf{K}) = \{\phi(a) \mid a \in \mathbf{K}\}$ *is an intermediate field. Show further that* $[\mathbf{K} : \mathbf{F}] = [\phi(\mathbf{K}) : \mathbf{F}]$.

$1 \in \phi(\mathbf{K})$, since $\phi(1) = 1$. $\phi(a) + \phi(b) \in \phi(\mathbf{K})$, since $\phi(a) + \phi(b) = \phi(a + b)$. Similarly for the other operations. Now, ϕ preserves linear independence over **F**, for

$$\sum_1^n c_i u_i = 0 \quad \text{if and only if} \quad \phi\left[\sum_1^n c_i u_i\right] = \sum_1^n c_i \phi(u_i) = 0,$$

where the c_i's $\in \mathbf{F}$ and the vectors $u_i \in \mathbf{K}$. Thus, a maximal linearly independent set in **K** over **F** corresponds to a maximal linearly independent set in $\phi(\mathbf{K})$ over **F**.

6. *Let* **F** *be a field. Numbers* $a_1, a_2, \ldots, a_n$ *are said to be algebraically independent over* **F** *if there is no nontrivial polynomial* $p(x_1, x_2, \ldots, x_n)$ *over* **F** *such that* $p(a_1, a_2, \ldots, a_n) = 0$. *Show that if* **F** *is countable, then for any n there exist n algebraically independent elements over* **F**.

If **F** is countable, we have seen (Problem 4, Section 1.5) that the set of all numbers algebraic over **F** is countable. Since **C** is not countable, we can pick a number a_1 which is transcendental over **F**. It is easy to see that $\mathbf{F}(a_1)$ is countable. By induction, then, we can construct a sequence $a_1, a_2, \ldots, a_n$ such that each a_k is transcendental over $\mathbf{F}(a_1, a_2, \ldots, a_{k-1})$. To show that these elements are algebraically independent over **F**, suppose there is a nontrivial relation $p(a_1, a_2, \ldots, a_n) = 0$. If k is the largest subscript appearing in a nonzero term, this implies that a_k is algebraic over $\mathbf{F}(a_1, a_2, \ldots, a_{k-1})$, a contradiction.

7. *Let $a_1, a_2, \ldots, a_n$ be algebraically independent over a field* $\mathbf{F}$. *Define* $\mathbf{E} = \mathbf{F}(a_1, a_2, \ldots, a_n)$. *Show that for every permutation of the numbers $a_1, a_2, \ldots, a_n$ there exists a unique $\phi \in G(\mathbf{E}/\mathbf{F})$ which has the effect of this permutation on the a_i's.*

Since any permutation can be obtained by a sequence of transpositions, we need only treat the case of a transposition. Furthermore, since the a_i's are algebraically independent over $\mathbf{F}$, it is easy to see that they can be adjoined to $\mathbf{F}$ in any order to obtain $\mathbf{E}$. Let us suppose that we are interested in the transposition $a_i \Leftrightarrow a_k$. Then we may write $\mathbf{E} = \mathbf{K}(a_i, a_j) = \mathbf{K}(c, d)$, where $c = a_i$, $d = a_j$, and $\mathbf{K}$ is $\mathbf{F}$ extended by all the other a's. An arbitrary element of $\mathbf{K}$ has the form $p(c, d)/q(c, d)$, where p and q are polynomials over $\mathbf{K}$. The obvious candidate for ϕ is the map

$$\phi : \frac{p(c, d)}{q(c, d)} \to \frac{p(d, c)}{q(d, c)}.$$

To show it is well defined, suppose $p_1(c, d)/q_1(c, d) = p_2(c, d)/q_2(c, d)$. By algebraic independence over $\mathbf{K}$, which is apparent, the polynomial $p_1q_2 - p_2q_1$ is identically 0. Thus $p_1(d, c)/q_1(d, c) = p_2(d, c)/q_2(d, c)$. The same reasoning may be used to show that ϕ is one-to-one. It is obvious that ϕ is onto and that it leaves $\mathbf{F}$ fixed; in fact, this ϕ leaves $\mathbf{K}$ fixed. Preservation of addition and multiplication is very simple, and so the details are omitted.

8. *How many automorphisms of* **R** *are there*?

There is only one, the identity map. For suppose ϕ is an automorphism of **R**. If $a < b$ then it must be that $\phi(a) < \phi(b)$, because $\phi(b) - \phi(a) = \phi(b-a) = \phi(\sqrt{b-a}\,)\cdot\phi(\sqrt{b-a}\,) > 0$. Also we know that any automorphism ϕ leaves each rational number fixed. If r is any real number, let a_n and b_n be sequences of rationals approaching r from below and above respectively. From $a_n < r < b_n$ we have $\phi(a_n) < \phi(r) < \phi(b_n)$ and so $a_n < \phi(r) < b_n$. Letting $n \to \infty$, we conclude that $\phi(r) = r$.

*9. *If the field of complex numbers* **C** *is an algebraic extension of a field* **E**, *show that any automorphism of* **E** *can be extended to an automorphism of* **C**.

This problem is more advanced than most; it is needed to solve Problem 10. They may both be skipped without loss of continuity.

Let ϕ denote the given automorphism of **E**. If **K** is any field, a mapping $\sigma : \mathbf{K} \to \mathbf{C}$ which is one-to-one and preserves the field operations is called an *isomorphism* of **K**. (If $\sigma(\mathbf{K}) = \mathbf{K}$ itself, then σ is also an automorphism of **K**.) Define a set S of ordered pairs $(\mathbf{K}, \sigma)$ such that **K** is a field containing **E** and σ is an isomorphism of **K** whose restriction to **E** is just ϕ. The set S can be partially ordered by the definition $(\mathbf{K}_1, \sigma_1) \leqslant (\mathbf{K}_2, \sigma_2)$ if and only if $\mathbf{K}_1 \subset \mathbf{K}_2$ and σ_2 restricted to $\mathbf{K}_1$ is just σ_1. Given two elements of S, they are not necessarily comparable under the relation "$\leqslant$". A subset of S such that all elements are comparable is called a *chain*, and every chain of S has an upper bound. To see this, denote the chain by $(\mathbf{K}_\alpha, \sigma_\alpha)$ for an index α. An upper bound is given by $(\mathbf{K}, \sigma)$ where $\mathbf{K} = \bigcup_\alpha \mathbf{K}_\alpha$ and $\sigma(a) = \sigma_\alpha(a)$ for any $\mathbf{K}_\alpha$ containing a. A famous principle called Zorn's Lemma asserts that if S is a partially ordered set in which every chain has an upper bound, then S has at least one maximal element (that is, an element s such that if $\bar{s} \geqslant s$, in fact $\bar{s} = s$). In the present case, this gives a maximal element which we shall call $(\overline{\mathbf{K}}, \bar{\sigma})$. (For Zorn's Lemma, see P. Halmos, *Naive Set Theory*. Zorn's Lemma is equivalent to the Axiom of Choice.)

We will now see that $\overline{\mathbf{K}} = \mathbf{C}$. If not, let r be a complex number not in $\overline{\mathbf{K}}$. Now r must be algebraic over $\overline{\mathbf{K}}$ (since $\mathbf{C}$ is an algebraic extension), and so it has a minimal polynomial over $\overline{\mathbf{K}}$. Under $\bar{\sigma}$, the coefficients of this polynomial are transformed so as to give another polynomial g, which is easily seen to be irreducible over $\bar{\sigma}(\overline{\mathbf{K}})$. Let t be any root of g. The mapping $r \to t$ induces an isomorphism of $\overline{\mathbf{K}}(r)$, the straightforward verification of which is omitted, and this is an extension of $\bar{\sigma}$. This construction contradicts the maximality of $(\overline{\mathbf{K}}, \bar{\sigma})$, thus implying that $\overline{\mathbf{K}} = \mathbf{C}$.

Since $\overline{\mathbf{K}} = \mathbf{C}$, it only remains to show that $\bar{\sigma}$ is onto, for then it must be an automorphism of $\mathbf{C}$. But if $t \in \mathbf{C}$, it has a minimal polynomial g over $\mathbf{E}$, the coefficients of which under ϕ^{-1} lead to another irreducible polynomial f over $\mathbf{E}$ of the same degree. Since $\bar{\sigma}$ is one-to-one and must map the roots of f to roots of g, it follows that t must be the image of one of them. Thus $\bar{\sigma}$ is onto, and this completes the solution.

*10. *How many automorphisms of* $\mathbf{C}$ *are there*?

There are an infinite number. To be more precise, let c be the cardinality of the continuum ($\mathbf{R}$, for example), and as is customary let 2^c be the cardinality of the set of all subsets of the continuum. Then there are 2^c automorphisms of $\mathbf{C}$. The proof depends partly on Problem 9 and also makes independent use of Zorn's Lemma, which was introduced in the solution to that problem. This result is included because it is quite striking, but the reader who lacks the appropriate background may skip it without any loss of continuity.

Let S be the family of all sets which are each algebraically independent over $\mathbf{Q}$, meaning strictly that each finite subset of each set is algebraically independent, according to the definition in Problem 6. If S is ordered by set inclusion, then each chain has an upper bound given simply by the union of the sets in the chain. Zorn's Lemma now implies that there is a maximal element G in S. G is called a *transcendence base* of $\mathbf{C}$ over $\mathbf{Q}$. Remember that it is a set of numbers algebraically independent over $\mathbf{Q}$. Let $\mathbf{E}$ be the smallest field containing all the elements of G. Clearly $\mathbf{C}$ is an algebraic extension of $\mathbf{E}$, or else G would not be maximal.

First we construct automorphisms of $\mathbf{E}$ and then we use Problem 9 to extend each to an automorphism of $\mathbf{C}$. Let f be a one-to-one mapping of G onto itself. Then f generates an automorphism of $\mathbf{E}$ in the following way. Every element of $\mathbf{E}$ has the form $p(\alpha, \beta, \dots)/q(\alpha, \beta, \dots)$, where p and q are polynomials over $\mathbf{Q}$ in some finite number of variables and $\alpha, \beta, \dots,$ denote some elements of G; that this is true follows since the set of all such expressions is itself a field and is a subset of any field containing all of G. The automorphism generated by f is the mapping.

$$\phi : \frac{p(\alpha, \beta, \dots)}{q(\alpha, \beta, \dots)} \to \frac{p(f(\alpha), f(\beta), \dots)}{q(f(\alpha), f(\beta), \dots)} .$$

It must be shown that ϕ is well defined, one-to-one, onto, and that it preserves the field operations. This depends on the algebraic independence of G and is quite straightforward. The details are essentially identical with those in the solution to Problem 7. By extending ϕ to all of **C**, as per Problem 9, we obtain from each f an automorphism of **C**.

Thus we now need to know how many such functions f there are. First observe that G must have cardinality c, for it is not hard to see that **E** has the same cardinality as G and also that **C** has the same cardinality as **E** (from the fact that **C** is algebraic over **E**). And of course the cardinality of **C** is c. Now the set of one-to-one mappings from a set of cardinality c to itself is 2^c. To see this let us look at such mappings from **R** to **R**. Let A be one of the 2^c subsets of the interval $(0, 1)$. Then $B = A \cup (-\infty, 0]$ and its complement D in **R** each have cardinality c. Thus there exist one-to-one maps from $(-\infty, 0)$ to B and from $[0, \infty)$ to D, the combination of which gives a one-to-one onto mapping from **R** to **R**. Since different sets A give rise to different mappings, we get 2^c mappings in this way. Thus there are at least 2^c one-to-one onto mappings f from G to G, and thus at least 2^c automorphisms of **C**. However, there can be no higher number of automorphisms of **C** because the total number of functions from **C** to **C** (not even necessarily one-to-one or onto) is still just 2^c. This can be seen by counting equivalently the functions from **R** to **R**, each of which uniquely determines a subset of $\mathbf{R}^2$ (or **C**) as its graph. So the number of such functions is no more than the number of subsets of **C**, which is 2^c.

Section 3.4

1. *If* $\mathbf{F}_1 = \mathbf{F}(r_1)$ *and* $\mathbf{F}_2 = \mathbf{F}(r_2)$ *and if* $\mathbf{F}_1$ *and* $\mathbf{F}_2$ *are conjugate fields over* **F**, *does it follow that* r_1 *and* r_2 *are conjugates over* **F**?

No. The number r_1 has a fixed finite number of conjugates. However $\mathbf{F}(r_2) = \mathbf{F}(qr_2)$ for every $q \in \mathbf{Q}$. Since there are an infinite number of these numbers qr_2, they cannot all be conjugates of r_1. Thus r_2 need not be a conjugate of r_1.

2. *If* **E** *is a finite extension of* **F**, *show that there are only a finite number of intermediate fields* **K**, $\mathbf{E} \supset \mathbf{K} \supset \mathbf{F}$. (*This already appeared as Problem* 12 *of Section* 2.2, *but an alternate and much simpler proof is now accessible.*)

First we need to extend **E** to a normal extension of **F**. If $\mathbf{E} = \mathbf{F}(r)$, let $\tilde{\mathbf{E}}$ be the splitting field of the minimal polynomial of r over **F**. Then $\tilde{\mathbf{E}} \supset \mathbf{E} \supset \mathbf{K} \supset \mathbf{F}$. But there are only a finite number of intermediate fields between $\tilde{\mathbf{E}}$ and **F**, for by Corollary 32, these are in one-to-one correspondence with the subgroups of $G(\tilde{\mathbf{E}}/\mathbf{F})$ which is finite and hence has only a finite number of subsets, let alone subgroups.

3. *Show that conjugacy is an equivalence relation on the set of subgroups of a group.*

Let G be the group and let H_1, H_2, and H_3 be subgroups of G. $H_1 = \phi^{-1}H_1\phi$ where $\phi = \text{id} \in G$; so the relation is reflexive. If $H_1 = \phi^{-1}H_2\phi$, then $H_2 = \phi H_1 \phi^{-1} = \psi^{-1}H_1\psi$ where $\psi = \phi^{-1} \in G$; so the relation is symmetric. If $H_1 = \phi^{-1}H_2\phi$ and $H_2 = \psi^{-1}H_3\psi$, then $H_1 = \phi^{-1}\psi^{-1}H_3\psi\phi = \tau^{-1}H_3\tau$ where $\tau = \psi\phi \in G$; thus the relation is transitive.

4. *Show that conjugacy is an equivalence relation on the set of finite extensions of a field* **F**.

Let $\mathbf{F}_1$, $\mathbf{F}_2$, and $\mathbf{F}_3$ be finite extensions of **F**. By the very definition, the relation is reflexive and symmetric. To show transitivity, the hypothesis is that $\mathbf{F}_1 = \mathbf{F}(r_1)$ and $\mathbf{F}_2 = \mathbf{F}(r_2)$, with r_1 and r_2 conjugates; also $\mathbf{F}_2 = \mathbf{F}(s_2)$ and $\mathbf{F}_3 = \mathbf{F}(s_3)$, with s_2 and s_3 conjugates. Let f and g denote the minimal polynomials (over $\mathbf{F}$) of the r_i's and the s_i's, respectively, and then let $\mathbf{E}$ be the splitting field of the product fg. Since $\mathbf{E}$ is normal over $\mathbf{F}$, Theorem 33 applies, thus enabling us to convert the question into one about groups. Since $G(\mathbf{E}/\mathbf{F}_1)$ and $G(\mathbf{E}/\mathbf{F}_2)$ are conjugate and also since $G(\mathbf{E}/\mathbf{F}_2)$ and $G(\mathbf{E}/\mathbf{F}_3)$ are conjugate, by the previous problem $G(\mathbf{E}/\mathbf{F}_1)$ and $G(\mathbf{E}/\mathbf{F}_3)$ are conjugate. By the theorem again, $\mathbf{F}_1$ and $\mathbf{F}_3$ are conjugate. (As an alternative, one could use Lemma 33.)

5. *If* $\mathbf{E}$ *is a normal extension of* $\mathbf{F}$ *and if* $\mathbf{K}$ *is an intermediate field, can every automorphism in* $G(\mathbf{K}/\mathbf{F})$ *be extended to an automorphism of* $\mathbf{E}$?

Yes. If $\mathbf{K} = \mathbf{F}(r)$, the given automorphism must map r to one of its conjugates r_j, and it is uniquely determined by this r_j. By the proof of Lemma 33, there is a $\phi \in G(\mathbf{E}/\mathbf{F})$ which does this.

6. *Let* $\mathbf{E}$ *be an extension of* $\mathbf{F}$, *and* ϕ *be an automorphism of* $\mathbf{F}$. *Can* ϕ *necessarily be extended to an automorphism of* $\mathbf{E}$: (*a*) *in general*? (*b*) *if* $\mathbf{E}$ *is a finite extension of* $\mathbf{F}$? (*c*) *if* $\mathbf{E}$ *is a normal extension of* $\mathbf{F}$?

The answer is negative in all three cases, as the following example shows. Let $\mathbf{F} = \mathbf{Q}(\sqrt{2})$ and $\mathbf{E} = \mathbf{F}(\sqrt[4]{2})$, where the roots refer to the unique positive roots. $\mathbf{E}$ is normal over $\mathbf{F}$ since it is the splitting field of $x^2 - \sqrt{2}$ over $\mathbf{F}$. (Of course $\mathbf{E}$ is not normal over $\mathbf{Q}$.) An automorphism of $\mathbf{F}$ is determined by the condition $\phi(\sqrt{2}) = -\sqrt{2}$, since $\sqrt{2}$ and $-\sqrt{2}$ are conjugates over $\mathbf{Q}$. Letting $s = \sqrt[4]{2}$, suppose ϕ could be extended to $\mathbf{E}$. We would then have: $\sqrt{2} = s^2$, $-\sqrt{2} = \phi(s^2) = \phi(s) \cdot \phi(s) = [\phi(s)]^2$, a contradiction.

7. *Let* r_1 *and* r_2 *be conjugates over a field* $\mathbf{F}$ *and suppose that* $r_2 \notin \mathbf{F}(r_1)$. *Prove or disprove*: $\mathbf{F}(r_1) \cap \mathbf{F}(r_2) = \mathbf{F}$.

The assertion is not necessarily true. First let us construct the general form of a counterexample and then a particular case. Let $\mathbf{F}$ be a finite extension of $\mathbf{Q}$ and let $\mathbf{K}_1$ be a nonnormal finite extension of $\mathbf{F}$. If we write $\mathbf{K}_1 = \mathbf{F}(s_1)$ for an appropriate s_1, then this s_1 must have some conjugate s_2 over $\mathbf{F}$ such that $s_2 \notin \mathbf{K}_1$. Otherwise $\mathbf{K}_1$ would be the splitting field of the minimal polynomial of s_1 over $\mathbf{F}$ and hence would be normal over $\mathbf{F}$. Now let $\mathbf{E}$ be the splitting field just mentioned, and define $\mathbf{K}_2 = \mathbf{F}(s_2)$. Since $\mathbf{K}_1$ and $\mathbf{K}_2$ are conjugate over $\mathbf{F}$, there exists a $\phi \in G(\mathbf{E}/\mathbf{F})$ such that $\phi(\mathbf{K}_1) = \mathbf{K}_2$. If it happens to be the case that $\mathbf{E}$ is also a normal extension of $\mathbf{Q}$, then since $\phi \in G(\mathbf{E}/\mathbf{Q})$ as well, it follows that $\mathbf{K}_1$ and $\mathbf{K}_2$ are conjugate over $\mathbf{Q}$. Thus we can write $\mathbf{K}_1 = \mathbf{Q}(r_1)$ and $\mathbf{K}_2 = \mathbf{Q}(r_2)$ for some conjugates r_1 and r_2. Nevertheless, $\mathbf{Q}(r_1) \cap \mathbf{Q}(r_2)$ contains all of $\mathbf{F}$ and hence more than just $\mathbf{Q}$.

The following particular example is of the type described. Let $\mathbf{F} = \mathbf{Q}(\sqrt{2})$, $\mathbf{K}_1 = \mathbf{F}(\sqrt[3]{2})$, and $\mathbf{K}_2 = \mathbf{F}(\omega \sqrt[3]{2})$, where ω is a primitive cube root of unity. Then $\mathbf{E} = \mathbf{F}(\sqrt[3]{2}, \omega)$ is normal over both $\mathbf{F}$ and $\mathbf{Q}$ since it is the splitting field of $x^3 - 2$ and $(x^3 - 2)(x^2 - 2)$ over $\mathbf{F}$ and $\mathbf{Q}$ respectively. Thus the required conditions are satisfied.

Section 3.5

1. *If* $\mathbf{E}$ *is a normal extension of* $\mathbf{F}$ *and* $\mathbf{K}_1$ *and* $\mathbf{K}_2$ *are intermediate fields, describe* $G(\mathbf{E}/\mathbf{K}_1 \cap \mathbf{K}_2)$.

It is the smallest subgroup of $G(\mathbf{E}/\mathbf{F})$ which contains both $G(\mathbf{E}/\mathbf{K}_1)$ and $G(\mathbf{E}/\mathbf{K}_2)$. Calling this subgroup H, we note that H is simply the set of all products of arbitrarily many automorphisms, some from $G(\mathbf{E}/\mathbf{K}_1)$ and the others from $G(\mathbf{E}/\mathbf{K}_2)$; for this set is a group and it must be contained in any group containing both $G(\mathbf{E}/\mathbf{K}_1)$ and $G(\mathbf{E}/\mathbf{K}_2)$. Thus, $\mathbf{K}_1 \cap \mathbf{K}_2$ is contained in the fixed field of H. On the other hand, if $a \notin \mathbf{K}_1 \cap \mathbf{K}_2$, it cannot be in the fixed field; for if $a \notin \mathbf{K}_1$, say, then there exists a $\phi \in G(\mathbf{E}/\mathbf{K}_1) \subset H$ such that $\phi(a) \neq a$, and similarly if $a \notin \mathbf{K}_2$. Since $\mathbf{E}$ is normal over $\mathbf{K}_1 \cap \mathbf{K}_2$, the fixed field of H, we have by Theorem 32 that $H = G(\mathbf{E}/\mathbf{K}_1 \cap \mathbf{K}_2)$.

2. *In the context of Problem* 1, *show that if further* $\mathbf{K}_1$ *and* $\mathbf{K}_2$ *are normal extensions of* $\mathbf{F}$, *then* $G(\mathbf{E}/\mathbf{K}_1 \cap \mathbf{K}_2) = \{\phi\psi \mid \phi \in G(\mathbf{E}/\mathbf{K}_1), \psi \in G(\mathbf{E}/\mathbf{K}_2)\}$.

The H discussed in the solution to Problem 1 must certainly contain the given set $S = \{\phi\psi \mid \phi \in G(\mathbf{E}/\mathbf{K}_1), \psi \in G(\mathbf{E}/\mathbf{K}_2)\}$. Also, S contains each of the groups $G(\mathbf{E}/\mathbf{K}_1)$ and $G(\mathbf{E}/\mathbf{K}_2)$. Thus it suffices to show that S is itself a group. By the fundamental theorem on Galois theory, for every $\tau \in G(\mathbf{E}/\mathbf{F})$, $\tau[G(\mathbf{E}/\mathbf{K}_i)] = [G(\mathbf{E}/\mathbf{K}_i)]\tau$, $i = 1, 2$. Thus, with the obvious notation, $(\phi\psi)^{-1} = \psi^{-1}\phi^{-1} = \phi'\psi' \in S$; $(\phi_1\psi_1)(\phi_2\psi_2) = \phi_1(\psi_1\phi_2)\psi_2 = \phi_1\phi_2'\psi_1'\psi_2 \in S$; and, of course, $\mathrm{id} = (\mathrm{id})(\mathrm{id}) \in S$. Thus S is a group.

3. *If* $\mathbf{E}$ *is a normal extension of* $\mathbf{F}$ *with intermediate fields* $\mathbf{K}_1$ *and* $\mathbf{K}_2$, *and if* $\mathbf{K}$ *is the smallest field containing* $\mathbf{K}_1$ *and* $\mathbf{K}_2$, *what is* $G(\mathbf{E}/\mathbf{K})$?

$G(\mathbf{E}/\mathbf{K}) = H_1 \cap H_2$, where $H_1 = G(\mathbf{E}/\mathbf{K}_1)$ and $H_2 = G(\mathbf{E}/\mathbf{K}_2)$. For if $\phi \in G(\mathbf{E}/\mathbf{K})$ then ϕ leaves fixed all elements of $\mathbf{K}$, which contains both $\mathbf{K}_1$ and $\mathbf{K}_2$. Thus $\phi \in H_1 \cap H_2$. Conversely, if $\phi \in H_1 \cap H_2$, then it leaves fixed the set of all rational combinations of elements of $\mathbf{K}_1$ and $\mathbf{K}_2$; but this set is easily seen to be a field, and hence it is $\mathbf{K}$ itself. Thus $\phi \in G(\mathbf{E}/\mathbf{K})$.

4. *Let* ω *be a primitive fifth root of unity. Find a field* $\mathbf{F}$ *such that* $[\mathbf{F}(\omega) : \mathbf{F}] = 2$.

Recalling the solution to Problem 1 of Section 2.7, we may take $\mathbf{F} = \mathbf{Q}(\sqrt{5})$.

5. *Let* ω *be a primitive* 31*st root of unity. Show that there exists a field* $\mathbf{F}$ *such that* $[\mathbf{F}(\omega) : \mathbf{F}] = 6$.

By Problem 1 of Section 3.3, $G(\mathbf{Q}(\omega)/\mathbf{Q})$ is cyclic of order 30. If ϕ generates this group, the subgroup generated by ϕ^5 has order 6. Letting $\mathbf{F}$ be its fixed field, by the normality of $\mathbf{Q}(\omega)$ over $\mathbf{Q}$ and hence over $\mathbf{F}$, we have $[\mathbf{Q}(\omega) : \mathbf{F}] = [\mathbf{F}(\omega) : \mathbf{F}] = |G(\mathbf{F}(\omega)/\mathbf{F})| = 6$.

6. *Find the order of the Galois group of* $x^4 - 2$ *and then describe completely, in some explicit way, each of its elements.*

If $\mathbf{E}$ is the splitting field of $x^4 - 2$, then $[\mathbf{E} : \mathbf{Q}] = 8$. For, $\mathbf{E} = \mathbf{Q}(i, \sqrt[4]{2})$ and $[\mathbf{E} : \mathbf{Q}] = [\mathbf{E} : \mathbf{Q}(\sqrt[4]{2})] \cdot [\mathbf{Q}(\sqrt[4]{2}) : \mathbf{Q}] = [\deg_{\mathbf{Q}(\sqrt[4]{2})} i][\deg_{\mathbf{Q}} \sqrt[4]{2}] = [2][4] = 8$. But then the group has order 8, since $\mathbf{E}$ is a normal extension of $\mathbf{Q}$. As in the proof of Theorem 19, a basis for the extension is given by $\{1, \sqrt[4]{2}, (\sqrt[4]{2})^2, (\sqrt[4]{2})^3, i, i\sqrt[4]{2}, i(\sqrt[4]{2})^2, i(\sqrt[4]{2})^3\}$. Thus an automorphism $\phi \in G(\mathbf{E}/\mathbf{Q})$ is completely described by its action on i and $\sqrt[4]{2}$. Since ϕ must preserve the roots of $x^2 + 1$ and $x^4 - 2$, we have $\phi(i) = \pm i$ and $\phi(\sqrt[4]{2}) = \pm\sqrt[4]{2}$ or $\pm i\sqrt[4]{2}$. All eight combinations yield the elements of the group. We may call these $\phi_1, \phi_2, \ldots, \phi_8$ and describe their behavior by the following table:

	ϕ_1	ϕ_2	ϕ_3	ϕ_4	ϕ_5	ϕ_6	ϕ_7	ϕ_8
$\sqrt[4]{2}$	$\sqrt[4]{2}$	$i\sqrt[4]{2}$	$-\sqrt[4]{2}$	$-i\sqrt[4]{2}$	$\sqrt[4]{2}$	$i\sqrt[4]{2}$	$-\sqrt[4]{2}$	$-i\sqrt[4]{2}$
i	i	i	i	i	$-i$	$-i$	$-i$	$-i$

7. *Find all the subgroups of the group obtained in the previous problem, and determine which ones are normal subgroups.*

By computation, we find that $\phi_2^2 = \phi_3$, $\phi_2^3 = \phi_4$, and $\phi_2^4 = \text{id} = \phi_1$. Also, $\phi_5^2 = \text{id}$, $\phi_6 = \phi_2\phi_5$, $\phi_7 = \phi_2^2\phi_5$, and $\phi_8 = \phi_2^3\phi_5$. Lastly, $\phi_5\phi_2 = \phi_2^3\phi_5$. Since everything can be expressed in terms of $\phi = \phi_2$ and $\psi = \phi_5$, we can rewrite the group as $\{\text{id}, \phi, \phi^2, \phi^3, \psi, \phi\psi, \phi^2\psi$, and $\phi^3\psi\}$, where $\phi^4 = \psi^2 = \text{id}$ and $\psi\phi = \phi^3\psi$.

Other than the entire group G and $\{\text{id}\}$, there are four normal subgroups: $N_1 = \{\text{id}, \phi, \phi^2, \phi^3\}$, $N_2 = \{\text{id}, \phi\psi, \phi^2, \phi^3\psi\}$, $N_3 = \{\text{id}, \psi, \phi^2, \phi^2\psi\}$, and $N_4 = \{\text{id}, \phi^2\}$. There are four nonnormal subgroups: $H_1 = \{\text{id}, \psi\}$, $H_2 = \{\text{id}, \phi\psi\}$, $H_3 = \{\text{id}, \phi^2\psi\}$, and $H_4 = \{\text{id}, \phi^3\psi\}$. (These answers are the results of a number of computations; but by Lagrange's theorem, only subsets with 2 or 4 elements need be tried.)

8. *If* **E** *is a normal extension of* **K** *and if* **K** *is a normal extension of* **F**, *does it follow that* **E** *is a normal extension of* **F**?

No. Let **E**, **K**, and **F** be the fixed fields of H_1, N_3, and G, respectively, in the solution to the previous problem. By the normality of H_1 in N_3 (as can be verified) and of N_3 in G, **E** is normal over **K** and **K** is normal over **F**. However, **E** is not normal over **F** because H_1 is not normal in G.

9. *If H is a normal subgroup of a group G and if N is a normal subgroup of H, must N be a normal subgroup of G?*

No, as shown in the solution to the previous problem.

10. *Why should it be obvious that the Galois group calculated in Problem* 6 *is solvable? Verify its solvability directly.*

Since $x^4 - 2$ is obviously solvable by radicals, its group must be solvable. The sequence $G \supset N_3 \supset H_1 \supset \{\text{id}\}$

fulfills the definition of solvability. Another such sequence is $G \supset N_1 \supset N_4 \supset \{\text{id}\}$.

11. *If* **F** *is a field containing no pth roots of the number* $A \in \mathbf{F}$, *for a fixed prime p, show that the polynomial* $x^p - A$ *is irreducible over* **F**.

Let ω be a primitive pth root of unity. By the first part of the proof of Lemma 34c, which does not depend on normality, $\mathbf{F}(\omega)$ contains no pth roots of A. For we may apply that proof with $\mathbf{K} = \mathbf{F}$ and $\tilde{\mathbf{K}} = \mathbf{F}(a)$ for each a, a pth root of A. Now, by the proof of Lemma 34h, $x^p - A$ is irreducible over $\mathbf{K} = \mathbf{F}(\omega)$, and so certainly it is irreducible over **F**.

Section 3.6

1. *Let* ω *be a primitive fifth root of unity, and label the roots of* $x^5 - 1$ *as*: $r_1 = 1$, $r_2 = \omega$, $r_3 = \omega^2$, $r_4 = \omega^3$, $r_5 = \omega^4$. *Give an example of a permutation of these roots which does not correspond to an element of the Galois group of the polynomial.*

The transposition (1 2) cannot be in the group, for, since $1 \in \mathbf{Q}$, 1 must be unchanged by every element of the group.

2. *Label the roots of* $x^4 - 2$ *as*: $r_1 = +\sqrt[4]{2}$ *(positive real)*, $r_2 = i\sqrt[4]{2}$, $r_3 = -\sqrt[4]{2}$, $r_4 = -i\sqrt[4]{2}$. *Give an example of a permutation of these roots which does not correspond to an element of the Galois group of the polynomial.*

The 3-cycle (1 3 4) cannot correspond to an element of the group, since if $\phi(r_1) = r_3$, then $\phi(r_3) = \phi(-r_1) = -r_3 = r_1 \neq r_4$.

3. *Prove or disprove: If* $f \in \mathbf{F}[x]$ *is irreducible, then its Galois group over* **F** *is transitive.*

The result is true, and the proof given in the first paragraph of the proof of Lemma 35a may be used verbatim with p replaced by n and $\mathbf{Q}$ replaced by $\mathbf{F}$. The hypothesis that p is prime was not used in that part of the proof.

4. *Find a fifth degree polynomial of the form $x^5 + bx + c$, such that it is not solvable by radicals.*

The polynomial $f(x) = x^5 - 5x + 5/2$ is such a polynomial. It has a positive local maximum at $x = -1$ and a negative local minimum at $x = +1$, and it has no other stationary points. Thus it has one real root on each of the intervals $(-\infty, -1)$, $(-1, 1)$, and $(1, +\infty)$. By Lemmas 35a and 35b, it is not solvable by radicals.

5. *If $f \in \mathbf{F}[x]$ is irreducible and there is a root r of f and a sequence of radical extensions $\mathbf{F} = \mathbf{F}_0 \subset \mathbf{F}_1 \subset \cdots \subset \mathbf{F}_N$ such that $r \in \mathbf{F}_N$, show that f is solvable by radicals over* $\mathbf{F}$. (*That is, if one root of an irreducible polynomial can be obtained by radicals, all the roots can.*)

The idea is to construct another sequence of radical extensions $\mathbf{F} = \mathbf{E}_0 \subset \mathbf{E}_1 \subset \cdots \subset \mathbf{E}_M$ such that $\mathbf{E}_M \supset \mathbf{F}_N$ and such that $\mathbf{E}_M$ is normal over $\mathbf{F}$. For then $\mathbf{E}_M$ will contain all the roots of f, by the definition of normality, and so f will be solvable by radicals.

For each j, we are given that $\mathbf{F}_{j+1} = \mathbf{F}_j(a_j)$, for some a_j such that $a_j^{n_j} \in \mathbf{F}_j$ for some n_j. Let us define $A_j = a_j^{n_j}$ and let the conjugates of A_j over $\mathbf{F}$ be $A_j, A_j', \ldots, A_j''$. Then the polynomial $g_j(x) = (x^{n_j} - A_j)(x^{n_j} - A_j') \cdots (x^{n_j} - A_j'')$ is readily seen to be over $\mathbf{F}$, for its coefficients are symmetric polynomials evaluated at the conjugates of A_j. Consider the polynomial $h(x) = \prod_{j=1}^{N-1} g_j(x)$, which is also over $\mathbf{F}$. Its splitting field $\mathbf{E}$ may obviously be obtained by a sequence of radical extensions $\mathbf{F} = \mathbf{E}_0 \subset \mathbf{E}_1 \subset \cdots$

$\subset \mathbf{E}_M = \mathbf{E}$. Moreover, $\mathbf{E}_M \supset \mathbf{F}_N$ and $\mathbf{E}_M$ is normal over $\mathbf{F}$. As noted earlier, this completes the proof.

6. *Suppose that the group of a polynomial f over a field $\mathbf{F}$ is S_n. Show that f is irreducible over $\mathbf{F}$. Use this result to show that if in addition $n \geqslant 5$, then no root of f may be expressed in radicals.*

If f were reducible over $\mathbf{F}$, then two of its roots would belong to different factors, and so they would not be conjugates. Thus the group of f over $\mathbf{F}$ would not contain any automorphisms mapping one of these to the other. Therefore the group could not contain all the permutations of the roots of f, and so it could not be S_n. If in addition $n \geqslant 5$, then the polynomial is not solvable by radicals over $\mathbf{F}$. By the previous problem, none of the roots may be expressed in radicals.

7. *Show that for every prime p, there exists a polynomial of degree p over $\mathbf{Q}$ which has S_p for its Galois group.*

By Lemma 35a, it suffices to find an irreducible polynomial of degree p over $\mathbf{Q}$ which has exactly two nonreal roots. For $p = 2$, $x^2 + 1$ suffices; for $p = 3$, $x^3 - 2$ suffices. Thus we may assume $p \geqslant 5$. Our construction only depends on the fact that p is odd.

Consider the function

$$g(x) = (x^2 + m)(x - 2)(x - 4) \cdots (x - 2(p - 2)),$$

where m is an even positive integer to be determined later. When x takes on the values $1, 3, 5, \ldots, 2p - 3$, the $p - 1$ values of $g(x)$ are integers greater than 2 in absolute value and alternating in sign. Thus the function $f(x) = g(x) - 2$ will also alternate signs at these $p - 1$ points. By the intermediate value theorem, f has at least $p - 2$ roots between 1 and $2p - 3$.

We saw in Problem 3 of Section 1.6 that the sum of the squares of the roots of a polynomial $\sum_{j=0}^{n} a_j x^j$ is

$$\left(\frac{a_{n-1}}{a_n}\right)^2 - \frac{2a_{n-2}}{a_n},$$

which does not depend on a_0. Thus this sum is the same for f as for g, since these polynomials only differ in the constant term. By considering g, we see that this sum is

$$2^2 + 4^2 + \cdots + [2(p-2)]^2 - m^2,$$

and so for m sufficiently large it is negative. Therefore at least one root of f is not real. But its complex conjugate must also be a root. Thus we know we have $p-2$ real roots and 2 nonreal roots, and these account for all p roots.

The resulting polynomial f is also irreducible, since it satisfies Eisenstein's criterion with the prime 2. (Note that 2^2 divides the constant term of $g(x)$, so it cannot divide the constant term of $f(x) = g(x) - 2$.)

8. *For every n, show that there exists a field* $\mathbf{F}$ *and an nth degree polynomial $f \in \mathbf{F}[x]$ such that the group of f over* $\mathbf{F}$ *is S_n. (Hint: Review Problems* 6 *and* 7 *of Section* 3.3.)

Choose numbers $a_1, a_2, \ldots, a_n$ that are algebraically independent over $\mathbf{Q}$. Let $\mathbf{E} = \mathbf{Q}(a_1, a_2, \ldots, a_n)$. Let $b_i = (-1)^i \sigma_i(a_1, a_2, \ldots, a_n)$, where σ_i is the ith elementary symmetric function. Define $\mathbf{K} = \mathbf{Q}(b_1, b_2, \ldots, b_n)$. Then we have $\mathbf{Q} \subset \mathbf{K} \subset \mathbf{E}$. Each permutation of the a_i's induces an automorphism of $\mathbf{E}$ (Problem 7, Section 3.3); and by the symmetry of the b_i's, such automorphisms leave $\mathbf{K}$ fixed. Thus $|G(\mathbf{E}/\mathbf{K})| \geqslant n!$. However, $\mathbf{E}$ is the splitting field over $\mathbf{K}$ of $f(x) = \prod_{i=1}^{n}(x - a_i)$. Therefore $|G(\mathbf{E}/\mathbf{K})| \leqslant n!$. We conclude that $|G(\mathbf{E}/\mathbf{K})| = n!$, and thus $G(\mathbf{E}/\mathbf{K}) = S_n$. $\mathbf{F} = \mathbf{K}$ satisfies the requirements of the problem. (The factor $(-1)^i$ in the definition of the b_i's was not necessary; it was introduced for consistency with later work.)

Section 3.7

1. *What is the group of an irreducible quadratic? Use the definition of a solvable group to show that it is a solvable group. Then carry through the procedure used in the first part of the proof of Galois' Theorem to derive an actual formula for its solution, which formula should turn out to be the quadratic formula.*

The group is S_2. For, if r_1 and r_2 are the roots of the quadratic, by Theorem 31 there is a unique ϕ in the Galois group such that $\phi(r_1) = r_2$; this is denoted (1 2). Thus, both (1 2) and of course the identity are in the group, and so the group is S_2. S_2 is solvable since $S_2 \supset \{\text{id}\}$ is a prime factor composition series, the single factor being 2. Now let us represent the quadratic as $f(x) = ax^2 + bx + c$. Our primitive square root of unity is just $\omega = -1$. As in the proof of Galois' Theorem, we define numbers a_0 and a_1:

$$a_0 = r_1 + r_2,$$
$$a_1 = r_1 - r_2.$$

We know that $a_0 = -b/a$; and by the proof cited, we know that a_1^2 should be in the coefficient field. By calculation we find that $a_1^2 = r_1^2 - 2r_1r_2 + r_2^2 = (r_1 + r_2)^2 - 4r_1r_2 = b^2/a^2 - 4c/a$. Consequently, $r_1 - r_2 = \pm\sqrt{b^2/a^2 - 4c/a} = (\pm\sqrt{b^2 - 4ac}\,)/a$. Combining this with $r_1 + r_2 = -b/a$, we obtain the usual quadratic formula.

2. *Repeat the previous problem for a cubic equation under the assumption that the group is S_3. Does your solution depend on the fact that the group is all of S_3? (Cf. Problem 1 of Section 3.1).*

Let the cubic be $f(x) = ax^3 + bx^2 + cx + d$ with roots r_1, r_2, and r_3. We have

$$a_0 = r_1 + r_2 + r_3$$

$$a_1 = r_1 + \omega r_2 + \omega^2 r_3$$

$$a_2 = r_1 + \omega^2 r_2 + \omega r_3$$

since $\omega^4 = \omega$ (for ω is a primitive cube root of unity). Let us take $\omega = (-1 + i\sqrt{3})/2$. From the proof of Galois' Theorem, a_1^3 and a_2^3 should be in the coefficient field. After an exhausting computation, we find that $a_1^3 = (-b/a)^3 + 9(bc/a^2 + 3d/a)/2 - 3i\sqrt{3\Delta^2/2}$, where $\Delta^2 = -4b^3d/a^4 - 27d^2/a^2 + 18bcd/a^3 - 4c^3/a^3 + b^2c^2/a^4$. ($\Delta^2$ is called the *discriminant* of the cubic; it equals $\Pi_{i<j}(r_i - r_j)^2$.) Similarly, $a_2^3 = (-b/a)^3 + 9(bc/a^2 + 3d/a)/2 + 3i\sqrt{3\Delta^2/2}$. Taking an arbitrary cube root in the first expression as the value of a_1 and computing a_2 from the relation $a_1a_2 = b^2/a^2 - 3c/a$, we finally are able to obtain:

$$r_1 = (-b/a + a_1 + a_2)/3$$

$$r_2 = (-b/a + \omega^2 a_1 + \omega a_2)/3$$

$$r_3 = (-b/a + \omega a_1 + \omega^2 a_2)/3.$$

The solution does not depend on the assumption that the group is S_3. (This assumption implies *a priori* that a_1^3 and a_2^3 are in the coefficient field, but the computation done here shows this is true for any cubic.)

3. *Let f be a cubic polynomial with real coefficients and let* **F** *be the smallest field containing the coefficients. If r_1, r_2, and r_3 denote the roots of f, we define the discriminant*

of f to be $\Delta^2 = \Pi_{i<j}(r_i - r_j)^2$. Show that $\Delta^2 \in \mathbf{F}$. Show further that if $\Delta^2 > 0$ all the roots are real and if $\Delta^2 < 0$ there is exactly one real root. What if $\Delta^2 = 0$?

$\Delta^2 \in \mathbf{F}$ since it is a symmetric function evaluated at the roots. If all the roots are real, it is clear that $\Delta^2 \geqslant 0$. Thus, if $\Delta^2 < 0$, there must be a nonreal root. Whenever there is a nonreal root, its complex conjugate must also be a root (Problem 11, Section 2.2), and so there must be exactly two nonreal roots, and hence exactly one real root. Let us denote such nonreal roots by $a \pm bi$, where $a, b \in \mathbf{R}$. Calling these roots r_1 and r_2, and the real root r_3, then $r_1 - r_2 = 2bi$. The product $(r_1 - r_3)(r_2 - r_3)$ is easily seen to be real, so that $\Delta^2 = -4b^2[(r_1 - r_3)(r_2 - r_3)]^2 < 0$. Thus, if $\Delta^2 > 0$, the roots must be real. By the same argument, if $\Delta^2 = 0$ the roots are real, and in this case at least two of them must be equal.

4. *If r is any root of the irreducible cubic f, over some field $\mathbf{F}$, and if Δ is a square root of the discriminant, show that the splitting field $\mathbf{E}$ of f, over $\mathbf{F}$, is given by $\mathbf{F}(\Delta, r)$.*

Since $\Delta = \pm(r_1 - r_2)(r_1 - r_3)(r_2 - r_3)$, where the r_i's are all the roots of f, it follows that $\mathbf{F}(\Delta, r) \subset \mathbf{E}$. Let us take $r_1 = r$. Since $(x - r_1)$ divides f in $\mathbf{F}(\Delta, r)[x]$, the quotient polynomial $(x - r_2)(x - r_3) = x^2 - (r_2 + r_3)x + r_2 r_3$ also has coefficients in the field $\mathbf{F}(\Delta, r)$. Substituting $x = r_1$, it follows that $(r_1 - r_2)(r_1 - r_3) \in \mathbf{F}(\Delta, r)$ and so, from the expression for Δ, $r_2 - r_3 \in \mathbf{F}(\Delta, r)$. Since also $r_2 + r_3 \in \mathbf{F}(\Delta, r)$, as it is the negative of the coefficient of a polynomial over this field, it follows that r_2 and r_3 both $\in \mathbf{F}(\Delta, r)$. Consequently $\mathbf{F}(\Delta, r) = \mathbf{E}$.

*5. *Let f be an irreducible cubic over $\mathbf{Q}$ with three real roots. Show that it is not possible to solve for any of its roots by real radicals alone. (This may be surprising.)*

We adopt the notation of the previous two solutions.

Suppose there were a sequence of real radical extensions, $\mathbf{Q} = \mathbf{F}_0 \subset \mathbf{F}_1 \subset \cdots \subset \mathbf{F}_N$, such that $\mathbf{F}_N$ contains a root of f. Then the additional real radical extension $\mathbf{F}_N(\Delta)$ would contain all the roots of f, by the result of the previous problem. As in the proof of Galois' Theorem, we may, without loss of generality, assume that each radical is of prime order. Since $\Delta^2 \in \mathbf{Q}$ (Problem 3), it is possible to reorder the radical extensions above so that $\mathbf{F}_1 = \mathbf{Q}(\Delta)$. Let $\mathbf{F}_M$ be the first field in the resulting list containing a root of f. By the previous problem, then, $\mathbf{F}_M$ contains all the roots of f. If $\mathbf{F}_M = \mathbf{F}_{M-1}(a)$ where $a^p \in \mathbf{F}_{M-1}$ and $a^p \notin \mathbf{F}_M$ (the nature of a radical extension of prime order), then since f has no roots in $\mathbf{F}_{M-1}$ and hence is irreducible over $\mathbf{F}_{M-1}$ (for if it were reducible, it would have a linear factor and hence a root), we must have $3 \mid p$ and so $3 = p$. The splitting field of f over $\mathbf{F}_{M-1}$ must be all of $\mathbf{F}_M$, because the fact that $[\mathbf{F}_M : \mathbf{F}_{M-1}]$ is prime precludes the existence of nontrivial intermediate fields. Therefore $\mathbf{F}_M$ is normal over $\mathbf{F}_{M-1}$. Since $x^3 - a^3$ has no roots in $\mathbf{F}_{M-1}$, not a by assumption and no others since they are not real, $x^3 - a^3$ is irreducible over $\mathbf{F}_{M-1}$. By definition of normality then, *all* the roots of $x^3 - a^3$ must be in $\mathbf{F}_M$, which contradicts the assumption that it is a real field.

6. *Use Galois' Theorem to show that every quartic is solvable by radicals.* (*Cf. Problem* 4, *Section* 3.1.)

S_4 is solvable because a prime power composition series can be shown to be $S_4 \supset A_4 \supset H_1 \supset H_2 \supset \{\text{id}\}$. Here A_4 is the set of odd permutations, $H_1 = \{\text{id}, (1\ 2)(3\ 4), (1\ 3)(2\ 4), \text{and } (1\ 4)(2\ 3)\}$, and $H_2 = \{\text{id}, (1\ 2)(3\ 4)\}$. The respective indexes of each subgroup in the previous one are 2, 3, 2, 2. Now if H is a subgroup of a group G and if the index of H in G is 2, it is easy to see that H must be normal. For if $g \in G$, then both gH and Hg are either both H or both the complement of H in G, depending on whether $g \in H$. Thus only the normality of H_1 in A_4 needs to be worked out, and this straightforward verification will be omitted. If H is any subgroup of S_4, then it can also be verified that $H \supset (H \cap A) \supset (H \cap H_1) \supset (H \cap H_2) \supset \{\mathrm{id}\}$ is a prime power composition series for H, where any repetitions are omitted. Thus any quartic has a solvable group, and is thus solvable by radicals. (It can be shown more generally that any subgroup of a solvable group is solvable.)

Section 4.1

1. *The function e^t is sometimes defined by the power series $\sum_{n=0}^{\infty} t^n/n!$. Show that this series converges for all t. Then use multiplication of series to prove the law of exponents $e^a e^b = e^{a+b}$.*

The series is absolutely convergent by the ratio test, since the ratio

$$\frac{|t|^{n+1}/(n+1)!}{|t|^n/n!} = |t|/(n+1)$$

approaches 0 as $n \to \infty$, for every t. For the law of exponents, we have:

$$\begin{aligned}
e^a e^b &= \left(\sum_{j=0}^{\infty} \frac{a^j}{j!}\right)\left(\sum_{k=0}^{\infty} \frac{b^k}{k!}\right) \\
&= \sum_{n=0}^{\infty} \sum_{j=0}^{n} \left(\frac{a^j}{j!} \cdot \frac{b^{n-j}}{(n-j)!}\right) \\
&= \sum_{n=0}^{\infty} \frac{1}{n!} \sum_{j=0}^{n} \frac{n!}{j!\,(n-j)!} a^j b^{n-j} \\
&= \sum_{n=0}^{\infty} \frac{1}{n!} (a+b)^n \\
&= e^{a+b}.
\end{aligned}$$

2. *On the basis of the fact that power series can be added term by term within their common radius of convergence, show that two series of the form* $f(t) = \sum_{k=-j}^{\infty} a_k t^k$ *and* $g(t) = \sum_{k=-j}^{\infty} b_k t^k$ *can also be added term by term. Here* $j > 0$ *and* f *and* g *both converge when* $0 < |t| < R$.

Since $t^j f(t)$ and $t^j g(t)$ are both convergent power series for $|t| < R$, they can be added term by term. Thus $t^j[f(t) + g(t)] = \sum_{k=-j}^{\infty}(a_k + b_k)t^{k+j}$. Division by t^j now gives the result.

3. *On the basis of the definition of series of the form* $\sum_{k=0}^{\infty} a_k t^k$ *in the case when the* a_k*'s may not be real, justify the fact that two such series may be multiplied in the usual way within their common radius of convergence.*

Let $\sum_{k=0}^{\infty} a_k t^k$ and $\sum_{j=0}^{\infty} A_j t^j$ be two such series. We write the real and imaginary parts of the coefficients as $a_k = b_k + ic_k$, $A_j = B_j + iC_j$. Within the common radius of convergence of the original two series, the four real series

$$\sum_{k=0}^{\infty} b_k t^k, \ \sum_{k=0}^{\infty} c_k t^k, \ \sum_{j=0}^{\infty} B_j t^j, \ \sum_{j=0}^{\infty} C_j t^j$$

all converge. We want to show that the product series is convergent and that the coefficient of t^n is $\sum_{k=0}^{n} a_k A_{n-k}$. If we compute the product, we obtain

$$\left[\sum_{k=0}^{\infty} b_k t^k + i \sum_{k=0}^{\infty} c_k t^k\right]\left[\sum_{j=0}^{\infty} B_j t^j + i \sum_{j=0}^{\infty} C_j t^j\right]$$

$$= \left[\sum_{k=0}^{\infty} b_k t^k \sum_{j=0}^{\infty} B_j t^j - \sum_{k=0}^{\infty} c_k t^k \sum_{j=0}^{\infty} C_j t^j\right]$$

$$+ i\left[\sum_{k=0}^{\infty} c_k t^k \sum_{j=0}^{\infty} B_j t^j + \sum_{k=0}^{\infty} b_k t^k \sum_{j=0}^{\infty} C_j t^j\right]$$

which is convergent by the rules for real series. Straightforward computation yields the expected coefficients.

4. *Prove Taylor's Theorem for the case of a polynomial: If f is a polynomial of degree n in y, then*

$$f(y) = f(y_0) + f^{(1)}(y_0)(y - y_0) + \frac{f^{(2)}(y_0)}{2!}(y - y_0)^2 + \cdots + \frac{f^{(n)}(y_0)}{n!}(y - y_0)^n.$$

Here y_0 is an arbitrary fixed value of y.

For $n = 0$, f is a constant and the result is obvious. If the statement is true for all polynomials of degree $< n$, then it is true for $f'(y)$. Thus we may write

$$f'(y) = f'(y_0) + f''(y_0)(y - y_0) + \cdots + \frac{f^{(n)}(y_0)}{(n-1)!}(y - y_0)^{n-1}.$$

Integration of both sides from y_0 to y now yields the result.

5. *Prove the following version of Gauss' Lemma: If $f(t, x)$ is a polynomial in two variables over $\mathbf{F}$ and if f can be written as a product $f(t, x) = r(t, x)s(t, x)$, where r and s are polynomials in x whose coefficients are rational functions of t over $\mathbf{F}$, then in fact f may be written $f(t, x) = \tilde{r}(t, x)\tilde{s}(t, x)$ where $\tilde{r}$ and $\tilde{s}$ are polynomials over $\mathbf{F}$ in two variables. Furthermore, $\tilde{r}$ and r have the same degree in x, as do $\tilde{s}$ and s.*

The proof of Lemma 12 applies almost verbatim, except that "± 1" should be replaced by "constant polynomial" and "prime" by "irreducible polynomial in t".

6. *Prove that the function* $\sqrt{1-t}$ *is analytic at* 0.

We seek a power series $\sum_{k=0}^{\infty} b_k t^k$ whose square is $1-t$. A necessary condition on the b_k's is obtained by squaring the proposed series and comparing coefficients with $1-t$. In particular,

$$\left(\sum_{k=0}^{\infty} b_k t^k\right)^2 = \sum_{k=0}^{\infty}\left[\sum_{j=0}^{k} b_j b_{k-j}\right] t^k = 1-t,$$

so that $b_0^2 = 1$, $2b_0 b_1 = -1$, and $2b_0 b_k = -\sum_{j=1}^{k-1} b_j b_{k-j}$. Clearly $b_0 = 1$ for the positive square root, hence $b_1 = -1/2$ and we have a recursion relation determining the b_k. If we can show that the resulting series has a positive radius of convergence, then we can conclude that it is a solution. By induction we can actually obtain a closed form expression for each b_k, $k \geqslant 3$, namely

$$-b_k = \frac{(2k-3)!}{2^{2k-2}k!(k-2)!} = \frac{3\cdot 5\cdot \cdots \cdot (2k-3)}{2^k k(k-1)!}$$

from which we conclude that $|b_k| < 1/4k$. The comparison

$$\sum_{k=3}^{\infty} |b_k t^k| < \sum_{k=3}^{\infty} \frac{1}{4k}|t|^k < \sum_{k=3}^{\infty} |t|^k$$

shows that our series certainly converges for $|t| < 1$. The addition of the first three terms cannot affect convergence.

An alternative approach is to apply Taylor's Theorem from calculus to $f(t) = \sqrt{1-t}$. This easily yields the coefficients b_k from the formula $b_k = f^{(k)}(0)/k!$, and thus we get a convergent series. The remainder term must be analyzed to show that the series converges to $f(t)$.

*7. *Let $f(t)$ be analytic at* 0 *and suppose* $f(0) > 0$. *Show that f has an analytic square root at* 0; *that is, show that there is a function $g(t)$ defined in some neighborhood of* 0 *such that $g(t)$ is analytic at* 0 *and* $[g(t)]^2 = f(t)$.

If $g(t)$ is one such function, then of course $-g(t)$ is another. Let us seek the positive square root, so that $g(0) > 0$. Define $x = g(t)$; we want to solve for x,

$$x^2 = a_0 + a_1 t + a_2 t^2 + \cdots,$$

in the form of a power series in t. By changing variables we can transform this problem so as to make it amenable to the argument used in the proof of Lemma 36b. Without loss of generality we may take $a_0 = 1$. The variable change $y = 1 - x$ then results in an equation

$$0 = c_1 t - y + c_0 y^2 + \sum_{k=2}^{\infty} c_k t^k,$$

where $c_0 = 1/2$ and $c_k = -a_k/2$ for $k \geqslant 1$. The exact method of the proof of Lemma 36b yields formulas for the coefficients in the proposed expansion $y = \sum_{k=1}^{\infty} b_k t^k$. To show convergence, a slight modification is necessary because there are an infinite number of the c_i's and we do not know they are bounded. But the convergence of $\sum_{k=0}^{\infty} a_k t^k$ obviously implies that of $\sum_{k=0}^{\infty} c_k t^k$, say absolutely on the set $|t| \leqslant r$. Thus as $k \to \infty$, $c_k r^k \to 0$, and so there is a constant A such that for all k, $|c_k| < A r^{-k}$. As in the proof of the lemma, replacement of each c_k by this bound yields a series for y, $y = \sum_{k=1}^{\infty} d_k t^k$, which dominates the original. But this series is the solution to the problem

$$0 = A\left(\frac{t}{r}\right) - y + A y^2 + A \sum_{k=2}^{\infty} \left(\frac{t}{r}\right)^k,$$

which is equivalent to

$$0 = Ay^2 - y + A\left(\frac{t/r}{1 - t/r}\right)$$

for $|t| < r$. From the quadratic formula, we see that y may be expressed as a power series in t/r, for $|t|$ sufficiently small, and hence as a power series in t. Thus $g(t) = x = 1 - y$ may also be so expressed.

8. *Consider the algebraic curve C defined by $f(t, x) = 0$. A real point of C is a point where both t and x are real. For example, the real points of $t^2 - x^2 = 0$ form two lines in the real t, x-plane. The curve $t^2 + x^2 = 0$ has only $(0, 0)$ for a real point. A real point is isolated if it has some neighborhood containing no other real points, as in the second example. Show that if $f(t, x)$ has real coefficients, then for every regular value of t, any corresponding real point cannot be isolated.*

In this case the coefficients of the power series for the root function, as derived in the proof of Lemma 36b, are all real. Thus the function is real and the points $(t, x(t))$ represent a real curve passing through the point in question.

9. *Suppose that $y = g(t)$ is a polynomial of degree m over $\mathbf{C}$. If there exist $m + 1$ values $t_i \in \mathbf{Q}$ at which $g(t_i) \in \mathbf{Q}$, show that the coefficients of $g(t)$ must actually be in $\mathbf{Q}$.*

The coefficients of $g(t)$ may be determined by a system of $m + 1$ equations as in the proof of Lemma 36c. The solution is obtained by rational operations on the values of t_i and $g(t_i)$. Thus in this case the coefficients are rational.

Section 4.2

1. *Verify that Kronecker's specialization gives a one-to-one correspondence between the sets P_d and K_d.*

There is a one-to-one correspondence between the ordered n-tuples $(i_1, i_2, \ldots, i_n)$, each i_j between 0 and $d-1$ inclusive, and the integers from 0 to $d^n - 1$. The n-tuple is simply the base d representation of the integer. From this we deduce a one-to-one correspondence between the monomials $\prod_{j=1}^{n} u_j^{i_j}$ and $y^{i_1 + i_2 d + \cdots + i_n d^{n-1}}$, and the extension to polynomials is immediate.

2. *If g, h, and gh are all in P_d, show that $\widehat{(gh)} = \hat{g}\hat{h}$.*

Let $h = a(u_0)\prod_{j=1}^{n} u_j^{i_j}$ and let $g = b(u_0)u_k$ for some k, $1 \leqslant k \leqslant n$. By the definition of $\hat{g}$ and $\hat{h}$, it is clear that the result holds in this case. By repeated application of this fact, the result holds for any monomials g and h. For any finite sum $\sum m_k$ of such monomials, it is clear that $\widehat{\sum m_k} = \sum \hat{m}_k$. Thus if we write $g = \sum m_k$ and $h = \sum n_j$, where the m_k's and n_j's are monomials, we obtain $\widehat{(gh)} = \widehat{(\sum_{k,j} m_k n_j)} = \sum_{k,j} \widehat{m_k n_j} = \sum_{k,j} \hat{m}_k \hat{n}_j = \hat{g}\hat{h}$. (This argument holds both when u_0 is a variable and when it is a constant.)

3. *In the example immediately preceding Lemma 36e, verify that the factorization $\hat{f}(y) = y^2(1 + y^4)$ leads back to a product gh which is not in P_d.*

Here recall that we are using $d = 3$. Therefore from $\hat{g}(y) = y^2$ we get $g(u_1, u_2) = u_1^2$; also from $\hat{h}(y) = 1 + y^4 = 1 + y \cdot y^3$, we get $h(u_1, u_2) = 1 + u_1 u_2$. The product $gh = u_1^2 + u_1^3 u_2$ is not in P_d because the exponent of u_1 in the second term is not less than $d = 3$.

4. *This problem refers to the proof of Theorem 36. Define $g_i(u_0, u_1, \ldots, u_n)$ by the requirement that $\hat{g}_i(u_0, y)$*

$= G_i(u_0, y)$. *Does it follow that* $q(u_0, u_1, \ldots, u_n) = \prod_{i \in S} g_i(u_0, u_1, \ldots, u_n)$?

Certainly not. This is essentially the point of the previous problem as well as the example in the text. In particular, $\prod_{i \in S} g_i(u_0, u_1, \ldots, u_n)$ may not be in P_d.

5. *Let* $f(t, x)$ *be an irreducible polynomial in two variables over* $\mathbf{Q}$, *and let* $\mathcal{H}$ *be the set of all rational numbers* α *such that* $f(\alpha, x)$ *is irreducible as a polynomial over* $\mathbf{Q}$ *in* x. $\mathcal{H}$ *is called a basic Hilbert set. Lemma 36d shows that* $\mathcal{H}$ *is infinite. Prove that* $\mathcal{H}$ *is dense in the real line* $\mathbf{R}$. (*A set* $\mathcal{S} \subset \mathbf{R}$ *is said to be dense in* $\mathbf{R}$ *if for every* $r \in \mathbf{R}$ *there exists a sequence of numbers* $s_n \in \mathcal{S}$ *converging to* r.)

The solution uses the properties of the changes of variables discussed at the beginning of the section, and so let us adopt the notation there. Let r be a fixed real number. For each n we want to define a number $s_n \in \mathcal{H}$, such that these s_n's converge to r.

Let n be fixed and choose t_0 to be a rational regular value of t such that $|t_0 - r| \leqslant 1/2n$. The corresponding polynomial $g(t, x)$ has precisely the properties to which the proof of Lemma 36d applies, and so there exist arbitrarily large rational values t_1 such that $g(t_1, x)$ is irreducible. If we pick any such $t_1 \geqslant 2n$, then $f(t_0 + (1/t_1), x)$ is irreducible. The number $s_n = t_0 + (1/t_1)$ is within distance $1/2n$ of t_0, hence within distance $\leqslant (1/2n) + (1/2n) = 1/n$ of r, and it belongs to $\mathcal{H}$. The resulting sequence of s_n's is therefore of the required type.

Section 4.3

1. *In the context of Lemma 37b, are there any automorphisms of* $\mathbf{E}$ *leaving* $\mathbf{F}$ *fixed other than those induced by the* $n!$ *permutations of the* a_i'*s*?

Yes, there are an infinite number of such automorphisms. Let r be any rational number other than 0 or 1, and define $\tilde{a}_1 = ra_1$. It is clear that $\tilde{a}_1, a_2, \ldots, a_n$ are algebraically independent over $\mathbf{F}$ and that $\mathbf{E}$ may also be written $\mathbf{E} = \mathbf{F}(\tilde{a}_1, a_2, \ldots, a_n)$. By Lemma 37b, any permutation of $\tilde{a}_1, a_2, \ldots, a_n$ yields an automorphism of $\mathbf{E}$ leaving $\mathbf{F}$ fixed. Furthermore, all permutations which are not the identity on $\tilde{a}_1$ are different from those arising from permutations of $a_1, a_2, \ldots, a_n$. Repeating for different values of r, we obtain an infinite number of such automorphisms. Of course, none of these automorphisms can leave $\mathbf{K}$ fixed, for the original automorphisms already provided the maximum number, $n!$, of elements of $G(\mathbf{E}/\mathbf{K})$.

2. *Are the elements $b_1, b_2, \ldots, b_n$, as defined in the text, algebraically independent over* $\mathbf{Q}$?

Yes. If we had a nontrivial polynomial $p(t_1, t_2, \ldots, t_n)$ over $\mathbf{Q}$ such that $p(b_1, b_2, \ldots, b_n) = 0$, then substitution for each t_i in terms of the s_j's, as defined in the text, would yield a nontrivial polynomial in the s_j's, say $q(s_1, s_2, \ldots, s_n)$. But then we would have $q(a_1, a_2, \ldots, a_n) = p(b_1, b_2, \ldots, b_n) = 0$, contradicting the algebraic independence of the a_i's.

3. *Let $H(t_1, t_2, \ldots, t_n, x)$ be a polynomial over* $\mathbf{Q}$ *in $n+1$ variables. Let $b_1, b_2, \ldots, b_n$ be n complex numbers which are algebraically independent over* $\mathbf{Q}$. *Define a polynomial $h(x) = H(b_1, b_2, \ldots, b_n, x)$ and consider it over* $\mathbf{K} = \mathbf{Q}(b_1, b_2, \ldots, b_n)$. *What is the relation between the irreducibility of H and the irreducibility of h?*

If h is irreducible, it does not follow that H is also irreducible, as shown by the example $H(t, x) = (t^2 + 1)(x^2 + 1)$ and any choice of a real transcendental number

b, such as π. However, if we add the hypothesis that the coefficient of the highest power of x in H is unity, then we can conclude that H is irreducible. This follows immediately from the proof of Lemma 37f.

On the other hand, if H is irreducible, then h is also irreducible. For if h is reducible, we know we can write

$$h(x) = [p_k x^k + \cdots + p_0][q_m x^m + \cdots + q_0]$$

where the p's and q's are rational functions of n variables over $\mathbf{Q}$, evaluated at the b_i's. By the argument used to develop Gauss' Lemma, we may assume that the p's and q's are actually polynomials over $\mathbf{Q}$. The coefficient of x^j in $h(x)$ is a polynomial d over $\mathbf{Q}$ evaluated at the b_i's. Thus we have

$$d(b_1, b_2, \ldots, b_n)$$
$$= \sum_{i=0}^{j} p_i(b_1, b_2, \ldots, b_n) q_{j-i}(b_1, b_2, \ldots, b_n).$$

Since the b_i's are algebraically independent, the polynomial $d - \sum_{k+m=j} p_k q_m$ must be identically 0. Since this holds for each j, we obtain the factorization

$$H = \left(\sum_{i=0}^{k} p_i x^i\right)\left(\sum_{i=0}^{m} q_i x^i\right).$$

4. *Show that none of the roots of the polynomial determined in the proof of Corollary 37b can be constructed.*

The polynomial is irreducible. By the solution to Problem 5 of Section 3.6, if one root could be obtained by a sequence of quadratic extensions of $\mathbf{Q}$, then every root could be so obtained. Therefore, if one root were constructible, all the roots would be constructible.

INDEX

(**Note:** Page references do not refer to every page on which a particular subject is mentioned, but generally only to the first page of a continuous discussion and to pages containing definitions or principal results.)